Introduction to Microwave Sample Preparation

Introduction to Microwave Sample Preparation

Theory and Practice

H. M. (Skip) Kingston, EDITOR
National Institute of Standards and Technology
(Formerly National Bureau of Standards)

L. B. Jassie, EDITOR
CEM Corporation

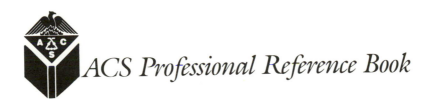

ACS Professional Reference Book

American Chemical Society, Washington, DC 1988

Library of Congress Cataloging-in-Publication Data

Introduction to microwave sample preparation.

(ACS professional reference book)
Includes bibliographies and indexes.
1. Chemistry, Analytic—Technique.   2. Microwave heating.   I. Kingston, H. M.,
1949–   . II. Jassie, L. B., 1940–   . III. Title: Microwave sample
preparation.   IV. Series.
QD75.4.S24I58   1988   543′.02   88–8139
ISBN 0–8412–1450–6

Copyright © 1988

American Chemical Society

PRINTED IN THE UNITED STATES OF AMERICA

Contents

About the Editors

H. M. (Skip) Kingston is a Research Chemist in the Inorganic Analytical Research Division at the National Bureau of Standards (NBS). He received his B.S. and M.S. degrees in chemistry from Indiana University of Pennsylvania in 1973 and 1975, and his Ph.D. in analytical chemistry from the American University in 1978. Various topics of research have included such diverse areas as nuclear waste stability and environmental water analysis and modeling. He began investigating microwave decomposition in 1979 as a method to more efficiently decompose Standard Reference Materials for certification. His current research interests include elemental separations, such as chelation and conventional chromatography, as well as laboratory robotics and automation, and future direction of the analytical laboratory. He has participated in over 80 elemental certifications at NBS.

L. B. Jassie is a Research Associate of the CEM Corporation at the National Bureau of Standards where she has been working since 1984 with H. M. (Skip) Kingston on Microwave Sample Preparation. Jassie received her B.S. in chemistry from Simmons College and an interdisciplinary M.S. in toxicology from the American University. She is presently a doctoral candidate in analytical chemistry with James F. Girard at the American University. She began her career in the Spectroscopy Laboratory of the Analytical Services Division of Bell Laboratories with Darwin Wood and followed that with a brief tenure as head of spectral services in the Analytical Lab at FMC Corporation. Current research interests are focused on the decomposition chemistry of biomolecules.

Contributors

Judith M. Babo   79
Department of Analytical Research
  and Development
The Squibb Institute for Medical
  Research
P.O. Box 191
New Brunswick, NJ 08903

Sador S. Black   79
Department of Analytical Research
  and Development
The Squibb Institute for Medical
  Research
P.O. Box 191
New Brunswick, NJ 08903

M. J. Collins   7
CEM Corporation
3100 Smith Farm Road
Matthews, NC 28105

L. B. Jassie   1, 93, 231
Inorganic Analytical Research
  Division
Center for Analytical Chemistry
National Bureau of Standards
Gaithersburg, MD 20899

H. M. Kingston   1, 93, 155, 231
Center for Analytical Chemistry
Inorganic Analytical Chemistry
  Division
National Bureau of Standards
Gaithersburg, MD 20899

John Labrecque   203
Kidd Creek Mines, Ltd.
Falconbridge Group Company
Analytical Laboratory, B–2002
Timmins, Ontario P4N 7K1,
  Canada

D. P. Manchester   187
CEM Corporation
3100 Smith Farm Road
Matthews, North Carolina 28105

S. A. Matthes   33
U.S. Department of the Interior
Bureau of Mines
1450 Queen Avenue, SW
Albany, OR 97321–2198

Z. Zakaria-Meehan   167
CEM Corporation
3100 Smith Farm Road
Matthews, NC 28105

E. D. Neas   7, 167
CEM Corporation
3100 Smith Farm Road
Matthews, NC 28105

K. Y. Patterson   155
Beltsville Human Nutrition
  Research Center
U.S. Department of Agriculture
Building 307, Room 226A
Beltsville, MD 20705

Patricia A. Stear 79
Department of Analytical Research
  and Development
The Squibb Institute for Medical
  Research
P.O. Box 191
New Brunswick, NJ 08903

E. F. Sturcken 187
E. I. du Pont de Nemours and
  Company
Savannah River Laboratory
Aiken, South Carolina 29808

T. S. Floyd 187
Floyd Associates
Route 6, Box 209
Clover, South Carolina 29710

C. Veillon 155
Beltsville Human Nutrition
  Research Center
U.S. Department of Agriculture
Building 307, Room 226A
Beltsville, MD 20705

R. Thomas White, Jr. 53
R. J. Reynolds Tobacco Company
Bowman Gray Technical Center
Winston–Salem, NC 27102

Foreword

_____ **Stuart A. Borman** _____

Sample dissolution is one of the most common operations in analytical chemistry. Because most quantitative analytical techniques require that samples be introduced in liquid form, thousands of sample dissolutions are performed every working day in chemical laboratories all over the world.

An IUPAC report on acid decomposition in trace element analysis[1] emphasizes the importance of this widely used sample preparation procedure:

> *The most common analytical techniques applied nowadays in trace element analysis of organic materials and minerals... normally begin with the dissolution of the substance to be analyzed.... Biological samples, as well as rocks, ores, slags, glass, etc., are rarely analyzed for trace elements without having received chemical pretreatment, during the course of which solid matter is brought into solution by decomposing and destroying the sample matrix.*

Despite the importance and widespread applicability of sample dissolution, most conventional digestion procedures are tediously labor-intensive, and a number of them, such as perchloric acid digestion, are potentially hazardous to laboratory personnel.

A number of the sample preparation procedures used today have been in use for more than 100 years. For example, heating samples in open beakers over flames or burners, a technique that doubtless predates the era of the alchemists, is still widely used today, especially when the modern hot plate is added to the list of applicable heating sources. However, the experimental conditions that prevail in open-beaker digestions are, at best, empirical.

From the age of alchemy we move on a few centuries to 1860, when G. L. Carius introduced the concept of closed-vessel sample dissolution. In a journal paper, Carius described the oxidation of a weighed quantity of sample with concentrated nitric acid in a sealed, strong-walled glass vial (Carius tube). As is the case with open-beaker digestions, Carius tube dissolutions continue to be widely used today, despite their hazards. The tubes must be heated, cooled, and then forcibly broken open in a protected area,

_____

[1]Jackwerth, E.; Gomiscek, S. *Pure Appl. Chem.* **1984,** 56(4), 479–489.

such as behind a plastic explosion shield. The tubes are also susceptible to unexpected explosions.

The continued presence in the laboratory of tedious and dangerous sample preparation techniques suggests the need for a more modern alternative. Fortunately, microwave dissolution, a promising alternative to the hot plate and Carius tube, has recently appeared on the horizon. The new microwave techniques described in this volume make it possible to speed the preparation of solid samples by combining the rapid heating ability of microwave energy with the advantages inherent in the use of sealed digestion vessels. Researchers have found microwave dissolution to be faster, more controlled, more elegant, and more amenable to automation than conventional open-beaker or closed-vessel techniques. According to Editor H. M. Kingston, "It is a new, more rapid tool for the analytical chemist involved in sample preparation, and it is capable of doing some preparations more reliably and accurately than can be done by conventional methods."

The advantages of microwave dissolution include faster reaction rates that result from the high temperatures and pressures attained inside the sealed containers. These containers are much more sophisticated than Carius tubes. They are generally made of polymers that will not contaminate or adsorb the sample and do not absorb microwave energy. The caps, which are screwed on to the canister with a torquing device, are designed to safely vent container gas in case of an excess internal pressure buildup.

The use of closed vessels also makes it possible to eliminate uncontrolled trace element losses of volatile molecular species that are present in a sample or that are formed in the course of a dissolution. Such losses can easily destroy the integrity of a measurement. It is well known in the field of elemental analysis[2] that significant percentages of elements such as arsenic, boron, chromium, mercury, antimony, selenium, and tin are lost at relatively mild temperatures with some open-vessel acid dissolution procedures. Several of these elements have already been shown to be retained when closed vessels are used.

Another advantage of microwave dissolution is a decrease in blank values as compared to open-beaker work, both because contamination from the laboratory environment is much lower and because closed vessels make it possible to use smaller quantities of reagents. The blank value incurred in sample preparation is a very important parameter because it frequently determines the limit of detection of a determination.

Since publication of their landmark article on microwave dissolution[3] and follow-up feature coverage of their work in a number of scientific journals, Kingston and Jassie have been inundated with phone calls and letters

---

[2] Hoffman, James I.; Lundell, G. E. F. *Journal of Research of the National Bureau of Standards,* **April 1939,** *22,* 465–470.
[3] Kingston, H. M.; Jassie, L. B. *Anal. Chem.* **1986,** *58,* 2534–2541.

from scientists all over the world who want to replace some of their flame, hot-plate, and furnace-based sample dissolutions and fusions with microwave procedures. For example, there was the toy-company chemist who wanted to use microwaves to dissolve toy samples so they could be tested for toxic constituents. There was the scientist responsible for 14 analytical laboratories who was "interested in any technique that may speed up the one million digestions of soil and rock we carried out in a year." There were literally hundreds of similar inquiries. In short, the interest in microwave dissolution has been tremendous and spontaneous.

As a result of a series of coordinated research efforts, microwave dissolution procedures have already been worked out for geological and metallurgical substances, botanical and food samples, clinical and biological specimens, petroleum products for Kjeldahl nitrogen determinations, and for many other important applications.

One of the most revolutionary aspects of microwave dissolution is the ease with which it can be automated. An example of this is the work of J. M. Labrecque of Kidd Creek Mines Ltd., who has interfaced a laboratory robot with an analytical balance, a cap-torquing device, and a microwave system. Labrecque is using this automated system to prepare the thousands of geological samples his laboratory handles each year. It is certain that robotic-controlled microwave dissolution will become increasingly popular as microwave and robotic instruments become more widespread.

In *Introduction to Microwave Sample Preparation: Theory and Practice*, readers will find most of what they need to know to get started in microwave dissolution: the history of the technique, the relevant methodological details, the all-important safety precautions, and a number of specific applications. The tremendous number of inquiries that Kingston and Jassie have received over the last several years strongly suggests that many researchers are indeed ready to relinquish their hot plates and perchloric acid fume hoods. If so, welcome to the revolution.

STUART A. BORMAN
Chemical and Engineering News
American Chemical Society
Washington, DC 20036

May 6, 1988

Preface

Solving real-world problems necessitates the return to basic science for understanding and theoretical knowledge. This pursuit of practical solutions is one of the driving forces behind research. Although phenomenal advances in analytical instrumentation have been made during the past several decades, techniques for the dissolution of sample matrices, one of the oldest and most persistent challenges encountered by the analytical chemist, had not changed markedly until the advent of microwave sample preparation.

Sample preparation is as responsible for accuracy and precision as is instrumental analysis. Many of the problems related to sample preparation are specific to individual sample matrices, analytes, and instrumental methods. Although panaceas are rare, the applicability of microwave-assisted sample preparation to all the major sample categories has met with a high degree of success.

In science and technology, it is important to disseminate information on a timely basis. Recent advances in analytical instrumentation and sample preparation techniques based on microwave technology, coupled with the keen interest expressed at analytical symposia, indicate that it is essential to compile and distribute the latest research in this rapidly progressing field. The scope of a Professional Reference Book must include fundamental concepts as well as the applications developed by leading researchers in the field.

This volume is intended for all chemists and analytical laboratory personnel who desire a thorough understanding of the theory and application of this new technology. The research described in this volume has established microwave sample preparation as a state-of-the-art technique. This reference book is designed to enable individuals to equip and operate a microwave sample preparation laboratory in a safe, productive, and cost-effective manner. We hope that it will stimulate appropriate use and will result in progress in the application of this technique.

The editors express their appreciation to the authors for their research and contributions to this first compendium of theories, techniques, and applications in microwave sample preparation.

H. M. (SKIP) KINGSTON
L. B. JASSIE
Inorganic Analytical Research Division
Center for Analytical Chemistry
National Bureau of Standards
Gaithersburg, MD 20899

June 3, 1988

Introduction to Microwave Acid Decomposition

——————————— L. B. Jassie and H. M. Kingston ———————————

"Every time a man puts a new idea across he finds ten men thought of it before he did—but they only thought of it".

Honoré Balzac

A chronological review of the literature on microwave-assisted sample preparation is presented. Temperature measurements and other innovations in the field of sample decomposition using microwave systems are surveyed.

CONVENTIONAL WET ASHING PROCEDURES involve heating a sample in acid for extended periods of time. In general, mineral acids, such as hydrochloric, nitric, phosphoric, sulfuric, and perchloric acids, digest materials more quickly at elevated temperatures and pressures (1). Classic acid digestions of samples for volatile or refractory elements require closed, chemically inert vessels such as Teflon PTFE [poly(tetrafluoroethylene)-] lined steel-jacketed bombs or glass Carius tubes that are heated in a conventional oven. These "bomb methods" are time-consuming, and can be expensive if Teflon liners and steel jackets need replacement, or if specially fabricated quartz tubes are required.

Early work in acid dissolution with microwave radiation as a heat source was accomplished by using home appliances that were often extensively modified to be used safely. Reaction vessels were placed inside evacuated desiccators or large plastic jars as a precautionary measure aimed at containment of the acid vapors and at reducing corrosion and the hazards of an explosion. When researchers realized that the small sample volumes in the cavity did not absorb all the magnetron power, additional loads were placed in the cavity to reduce reflected radiation, which damages the magnetron and alters its power output. These auxiliary loads reduce microwave power that would be absorbed by acids and samples; consequently, the samples receive neither constant nor reproducible power, a situation that frequently produces variable or incomplete dissolutions.

1450–6/88/0001$06.00/0

The use of microwave energy as a heat source in wet ashing procedures was first demonstrated in 1975 (2). Most of the early papers described specific applications using open or covered vessels (at atmospheric pressure) for the acid dissolution of bone (3), biological tissue (4–6), and botanical matrices (7). Several studies (8–10) compared the technique with different digestion procedures, and examples have been given (11–13) of this technique's successful application to biological samples suitable for atomic absorption and emission spectroscopy. Early researchers knew that open-vessel work involved the risk of environmental contamination as well as mechanical or volatile loss of the analyte. Open-vessel conditions also limited the maximum sample temperature to the boiling point of the acids. In an attempt to deal with these problems, investigators turned to closed polycarbonate bottles (14, 15) and Teflon PFA [(Perfluoro alkoxy)ethylene] digestion vessels to obtain the high temperatures and pressures needed for the digestion of steels (16); geologic species such as ores, zircons and other rocks (17–19) including soils (20); environmental samples (21); and biological samples (22–24).

Not only were significant reductions in sample preparation time realized as the result of the high temperatures and pressures that could be obtained in 2–3 min, but new applications for microwave heating in closed containers became apparent. For instance, the elevated pressure in closed reaction vessels heated by microwaves increased not only the reaction rate in organic syntheses, but the product yield as well (25).

Although wet ashing of samples with microwaves was described over a decade ago, the method remained something of a curiosity. Since 1984, however, there has been renewed interest in microwave-based sample preparation for analytical chemistry. The results of research on the extraction of metals from sediments (26) and research on trace elements in biological tissue (27–29) have been presented at various regional and national conferences. The first conference session devoted solely to presentations in the field of sample preparation using microwave dissolution took place at the 1986 Eastern Analytical Symposium in New York City. In 1986, four articles in *Analytical Chemistry* were devoted to the subject of microwave dissolution and its inherent suitability for steels (16), geologic materials (17–19), and biological matrices (23). Conference presentations and journal articles have indicated keen interest in recovery studies (30–32), the chemistry of decomposition in biological matrices (33–35), and mineralization of blood in a flow injection system (36). All efforts have demonstrated that this dissolution technique gives results comparable with those obtained by classic methods. Only after the first serious attempts were made to investigate and understand the various parameters that influence acid dissolution in microwave systems (23) was attention given to temperature and pressure conditions during digestion.

Temperature changes resulting from microwave heating have been measured with thermocouples to determine the kinetic parameters of both biochemical and other chemical reactions (37). Thermocouple devices have been used successfully to determine both the thermodynamic functions of chemical reactions in aqueous solutions (38) and thermal hemolytic thresholds of erythrocytes in saline solutions that have been heated by microwave energy (39). Although useful for a wide range of temperatures, thermocouples are somewhat difficult to construct in the narrow, shielded configurations required for acid dissolution in microwave systems (23). Recently, fiber-optic thermometry has been used to measure microwave-induced hyperthermia (40), in the fabrication of composite materials cured with microwave power (41), and in monitoring acid-initiated sample decomposition in a microwave system (42). Localized heating (43) and electrical interference problems common to metal probes (44,45) do not occur in fiber-optic temperature sensors (42).

The relationship between the acid used and microwave interaction resulting in sample digestions has, until very recently, remained purely empirical. Until the obstacles associated with temperature and pressure measurement in closed-vessel microwave digestions were overcome, this parameter could not be studied in a systematic fashion and, without temperature and pressure measurements, the principles involved in this process could not be investigated. Control of the temperature and pressure during closed vessel work is critical to the efficiency, reproducibility, and above all, the safety of the procedure (23). Continuous, real-time monitoring of the temperature provides new insights into matrix component decompositions (34, 42). This capability allows the decomposition mechanism to be observed and general procedures to be applied to a variety of sample types. Current research is aimed at demonstrating that samples digested by using microwave techniques can produce results that are equivalent to, or better than, traditional digestion procedures for the analysis of certain elements, and also at firmly establishing and quantifying the theoretical basis for understanding the technique.

Several factors have contributed significantly to the proliferation of effective sample preparation techniques using microwave dissolution, and to the research efforts aimed at providing the basis for understanding how it works. The development of specially designed equipment for analytical chemical use has been most important. Commercial analytical microwave systems address, for the first time, problems such as acid fumes, sample power reflection, field inhomogeneity, and long-duty cycles that were encountered by analysts trying to modify home appliances. Until a strong, inert, and microwave-transparent container for acid dissolution was fabricated from Teflon PFA, closed-vessel experimentation was dangerous and limited. Finally, routine measurement of the elevated dissolution temperatures and pressures during microwave exposure was difficult until the equipment modifications, Teflon vessels, and new fiber optic thermometry were applied simultaneously. The evolution of microwave systems for the laboratory has provided researchers with new and better tools to irradiate, contain, and measure the dissolution process in a safe and reliable fashion.

The theoretical basis for understanding microwave dissolution lies in an examination of related topics in classic physical chemistry, thermodynamics, electromagnetic radiation, and dielectric materials. The ability to apply simplified thermodynamic relationships to actual measurements enables the analyst to gain the practical understanding that allows the technique to be generalized to a particular sample matrix. This ability to generalize from concepts to real samples makes this technique useful to the majority of the analytical community preparing samples for today's modern instrumentation.

Literature Cited

1. Jackwerth, E.; Gomiscek, S. *Pure Appl. Chem.* **1984**, 56(4), 480–489.
2. Abu-Samra, A.; Morris, J. S.; Koirtyohann, S. R. *Anal. Chem.* **1975**, 47, 1475–1477.
3. Brown, A. B.; Keyzer, H. *Contrib. Geol.* **1978**, 16, 85–87.
4. Barrett, P.; Davidowski, Jr, L. J.; Penaro, K. W.; Copeland, T. R. *Anal. Chem.* **1978,** 7, 1021–1023.
5. Cooley, T. N.; Martin, D. F.; Quincel, R. H. *J. Environ. Sci. Health, Part A* **1977,** 12(1&2), 15–19.
6. Andoh, K.; Saitoh, T.; Takatani, A.; Takahashi, F.; Tazuya, Y.; Tsunajima, K.; Motoki, C.; Yasuoka, K.; Yamaji, Y.; Natsuoka, C. *Kenkyu Kiyo-Tokushima Bunri Daigaku* **1982**, 25, 113–125. *Chem. Abstr.* **1982**, 97, 125965c.
7. Keyzer, H. *Chemistry in Australia* **1978**, 45(2), 44.
8. White, R. T.; Douthit, G. E. *J. Assoc. Off. Anal. Chem.* **1985**, 68(4), 766–769.
9. Nadkarni, R. A. *Anal. Chem.* **1984,** 56, 2233–2237.
10. De Boer, J. L. M.; Maessen, F. J. M. *J. Spectrochim. Acta, Part B* **1983**, 38, 379–746.
11. Matsumura, S.; Karai, I.; Takise, S.; Kiyota, I.; Shinagawa, K.; Horiguchi, S. *Osaka City Med. J.* **1982**, 28, 145–148.
12. Blust, R.; Van der Linden, A.; Decleir, W. *Atomic Spectroscopy* **1985**, 6(6), 163–165.
13. DeMenna, G. J. Presented at the 23rd Eastern Analytical Symposium, New York, 1984, Paper No. 166.
14. Matthes, S. A.; Farrell, R. F.; Mackie, A. *J. Tech. Prog. Rep.–US, Bur. Mines* **1983**, 120.
15. Lamothe, P. J.; Fires, T. L.; Consul, J. J. *Anal. Chem.* **1986**, 58, 1881–1886.
16. Fernando, L. A.; Heavner, W. D.; Gavrielli, C. C. *Anal. Chem.* **1986**, 58, 511–512.
17. Fischer, L. B. *Anal. Chem.* **1986,** 58, 261–263.
18. Westbrook, W. T.; Jefferson, R. H. *J. Microwave Power* **1986**, 21(1), 25–32.

19. Smith, F.; Cousins, B. *Anal. Chim. Acta* **1985**, *177*, 243–245.
20. Papp, C. S. E.; Fischer, L. B. *Analyst* **1987**, *112*, 37–338.
21. Bettinelli, M.; Baroni, U.; Pastorelli, N. *J. Anal. At. Spectr.* **1987**, *2(5)*, 485–489.
22. Jassie, L. B.; Kingston, H. M. 1985 Pittsburgh Conference Abstracts, Paper 108A.
23. Kingston, H. M.; Jassie, L. B. *Anal. Chem.* **1986**, *58*, 2534–2541.
24. Nakashima, S.; Sturgeon, R.; Willie, S.; Berman, S. *Analyst* **1988**, *113*, 159–163.
25. Gedye, R.; Smith, F.; Westaway, K.; Ali, H.; Baldisera, L.; Laberge, L.; Rousell, J. *Tetrahedron Lett.* **1986**, *27(3)*, 279–282.
26. Mahan, K. I.; Foderaro, T. A.; Garza, T. L.; Martinez, R. M.; Maroney, G. A.; Trivisonno, M. R.; Willging, E. M. *Anal. Chem.* **1987**, *59*, 938–945.
27. Kingston, H. M.; Jassie, L. B.; Fassett, J. D., Presented at the 190th National Meeting of the American Chemical Society, Chicago, IL, September 1985, Paper No. ANAL 10.
28. Veillon, C.; Patterson, K. Y.; Kingston, H. M. Presented at the 28th Rocky Mountain Conference, Denver, CO, August, 1986, Paper No. 8.
29. Veillon, C.; Patterson, K. Y.; Kingston, H.M., Presented at the 13th FACSS, St. Louis, MO, September, 1986, Paper No. 689.
30. Revesz, R.; Hasty, E. Presented at the 38th Pittsburgh Conference and Exposition, March, 1987, Atlantic City, NJ, Paper No. 252.
31. Copeland, T. Presented at the 2nd Annual U.S. EPA Symposium on Solid Waste Testing and Quality Assurance, July, 1986, Washington, DC.
32. Binstock, D. A.; Grohse, P. M.; Swift, P. L.; Gaskill, A.; Copeland, T. R.; Friedman, P. H. Presented at the 3rd Annual Symposium on Solid Waste Testing and Quality Assurance, July, 1987, Wash. DC Paper No. 5–1.
33. Kingston, H. M; Jassie, L. B. Presented at the 3rd Annual U.S. EPA Symposium on Solid Waste Testing and Quality Assurance, July, 1987, Washington, DC.
34. Kingston, H. M.; Jassie, L. B. *J. Res. Natl. Stds.* **1988,** *93(3)*, 269–274 Accuracy in Trace Analysis Symposium Proceedings, September, 1987, Gaithersburg, MD.
35. Pratt, K. W.; Kingston, H. M.; MacCrehan, W. A.; Koch, W. F. *Anal. Chem.* **1988** in press.
36. Burguera, M.; Burguera, J. L. *Anal. Chim. Acta* **1986**, *179*, 351–357.
37. Bacci, M.; Bini, M.; Checcucci, A.; Ignesti, A.; Millanta, L.; Rubino, N.; Vanni, R. *Proc. 14th Microwave Power Symp.* Monaco, 1979, p 42–44.
38. Bacci, M.; Bini, M.; Checcucci, A.; Ignesti, A.; Millanta, L.; Rubino, N.; Vanni, R. *J. Chem. Soc. Faraday Trans. 1* **1981**, *77*, 1503–1509.
39. Checcucci, A.; Olmi, R.; Vanni, R. *J. Microwave Power* **1985**, *20(3)*, 161–163.
40. Papoutis, D. *Photonics Spectra* **1984**, *March*, 5360.
41. Wickersheim, K. A.; Sun, M. H. *J. Microwave Power* **1987**, *22(2)*, 85–93.
42. Kingston, H. M.; Jassie, L. B. Presented at the 25th Eastern Analytical Symposium, October, 1986, Paper No. 76.

43. Bini, M.; Ignesti, A.; Olmi, Rubino, N.; Vanni, R.; Millanta, L. Abstracts of Papers, XXIst General Assembly URSI, 1983, Paper No. 8.8.
44. Baker, R. J.; Smith, V.; Phillips, T. L.; Kane, L. J.; Kobe, L. H. *IEEE Transaction, Microwave Theory and Techniques* **1978,** MTT–26(*No.* 8), 541–545.
45. Olmi, R., IROE–CNR, Firenze, Italy, personal communication, 1986.

RECEIVED for review January 25, 1988. ACCEPTED revised manuscript May 6, 1988.

CHAPTER 2

Microwave Heating

Theoretical Concepts and Equipment Design

E. D. Neas and M. J. Collins

"Science is built up with facts, as a house is with stones. But a collection of facts is no more a science than a heap of stones is a house".

Jules Henri Poincaré

This chapter discusses the theoretical concepts of dielectric loss, ionic conduction, dipole rotation, and sample size as they relate to microwave heating for acid dissolutions. The chapter includes the design of microwave equipment and accessories to meet the heating requirements for acid dissolution. These microwave instruments protect the magnetron, prevent corrosion, and provide uniform heating. The accessories include digestion vessels, turnable systems, and microwave furnaces.

THE DEVELOPMENT OF RADAR DURING WORLD WAR II stimulated the rapid growth of microwave technology. The first microwave heating applications soon followed and included heating food with microwave energy (*1*). This development led to the large-scale use of domestic microwave ovens. Investigation of industrial applications of microwave heating also began in the 1940s. Industrial applications of microwave technology include the treatment of coal with microwaves to remove organic sulfur and other potential pollutants (*2*), frozen food tempering, rubber vulcanization, and pasta product drying (*3*).

One successful analytical development has been the use of a microwave drying system for determining moisture and crude fat in meat and poultry products (*4–7*). Because of the need for a better understanding of how microwaves interact with different samples (applications development) and the need for proper hardware, the acceptance of microwave technology for rapid heating and drying in analytical applications has been very slow. Analytical chemists are very familiar with conductive heating (e.g. heating with hot plates, convection ovens, or flame). Microwave heating involves direct absorption of energy by the material to be heated. Therefore, new methods are required to properly apply this technology to analysis.

1450–6/88/0007$07.50/0

Until recently, the proper hardware for microwave drying that would allow the analytical chemist to fully use the technology and heating had not been available. Many laboratories have tried to use appliance grade microwave ovens. In most cases, chemists soon discovered that the domestic microwave oven was inadequate for the chemical analysis. The appliance grade microwave ovens were designed to cook fairly large food samples by heating them. Analytical samples tend to be much smaller, may be volatile, can give off toxic fumes, can lose valuable elements during the heating process, etc. These sample characteristics place unique and stringent requirements on the microwave heating system that conventional domestic microwave ovens cannot satisfy.

Most of the chapters in this book discuss the progress that has been made in applications development using microwave energy for acid dissolutions. This chapter discusses the theory of microwave heating and the design of microwave equipment to meet the heating requirements for acid dissolution.

Theory

Microwaves are electromagnetic energy. Microwave energy is a nonionizing radiation that causes molecular motion by migration of ions and rotation of dipoles, but does not cause changes in molecular structure. Microwave energy has a frequency range from 300 to 300,000 MHz (Figure 2.1). Four fre-

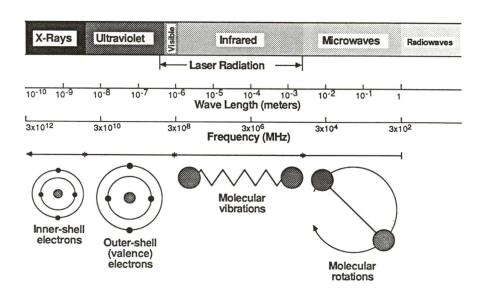

Figure 2.1. Electromagnetic spectrum.

quencies are used for industrial and scientific microwave heating and drying: 915 ± 25, 2450 ± 13, 5800 ± 75, and $22,125 \pm 125$ MHz. These frequencies were established for industrial, scientific, and medical use by the Federal Communications Commission and conform to the International Radio Regulations adopted at Geneva in 1959. Of these frequencies, 2450 MHz is the most commonly used, and is the frequency used in all home microwave units. The typical energy output in a microwave system is 600–700 W. Thus, within 5 min, approximately 43,000 cal is supplied to the microwave cavity for sample heating.

Dielectric Loss

The heating pattern of a sample that is heated with microwave energy will depend, in part, upon the dissipation factor of the sample ($\tan \delta$). The dissipation factor is a ratio of the sample's dielectric loss or "loss" factor (ϵ'') to its dielectric constant (ϵ'); $\tan \delta = \epsilon''/\epsilon'$. The dielectric constant is a measure of a sample's ability to obstruct the microwave energy as it passes through, and the loss factor measures the sample's ability to dissipate that energy (8). The word "loss" is used to indicate the amount of input microwave energy that is lost to the sample by being dissipated as heat. Tables with dissipation factors and dielectric constants are available for different materials (9).

When microwave energy penetrates a sample, the energy is absorbed by the sample at a rate dependent upon its dissipation factor. Penetration is considered infinite in materials that are transparent to microwave energy, and is considered zero in reflective materials, such as metals. The dissipation factor is a finite amount for absorptive samples. Because the energy is quickly absorbed and dissipated as microwaves pass into the sample, the greater the dissipation factor of a sample, the less the penetration of the microwave energy at a given frequency. A useful way to characterize penetration is by the half-power depth for a given sample at a given frequency. The half-power depth is that distance from the surface of a sample at which the power density is reduced to one-half that at the surface (10). The half-power depth varies with the dielectric properties of the sample and approximately with the inverse of the square root of the frequency.

Typically, microwave energy is lost to the sample by two mechanisms: ionic conduction and dipole rotation. In many practical applications of microwave heating, ionic conduction and dipole rotation take place simultaneously.

Ionic Conduction

Ionic conduction is the conductive (i.e., electrophoretic) migration of dissolved ions in the applied electromagnetic field. This ionic migration is a

Table 2-1. Effect of Increasing NaCl Concentration
on the Dissipation Factor

Molal Concentration	Tangent[a] ∂ ($\times$ 10)
0.0[b]	1570
0.1	2400
0.3	4350
0.5	6250

[a]Measurements made at 3000 MHz and 25 °C. Data extracted from reference 9.
[b]Water only.

flow of current that results in I^2R losses (heat production) due to resistance to ion flow. All ions in a solution contribute to the conduction process, but the fraction of current carried by any given species is determined by its relative concentration and its inherent mobility in the medium. Therefore, the losses due to ionic migration depend on the size, charge, and conductivity of the dissolved ions, and are subject to the effects of ion interaction with the solvent molecules (11).

The parameters affecting ionic conduction are ion concentration, ion mobility, and solution temperature. Every ionic solution will have at least two ionic species (e.g. Na^+ and Cl^- ions) and each species will conduct current according to its concentration and mobility. Table 2-1 shows that an increase in ion concentration will increase the dissipation factor. The contribution of ionic conductance to microwave heating is illustrated in Table 2-1 by the large increase in the dissipation factor when NaCl is added to water. The dissipation factor of an ionic solution will change with temperature because temperature affects ion mobility and concentration.

Dipole Rotation

Dipole rotation refers to the alignment, due to the electric field, of molecules in the sample that have permanent or induced dipole moments. Dipole rotation is illustrated in Figure 2.2. As the electric field of the microwave energy increases, it aligns the polarized molecules (Figure 2.2a). As the field decreases, thermally induced disorder is restored (Figure 2.2b). Figures 2.2a and 2.2b are grossly exaggerated for the sake of illustration. In fact, the applied microwave field causes the molecules, on average, to temporarily spend very slightly more time pointing in one direction rather than in other directions. Associated with that tiny bit of preferred orientation there is a tiny bit of molecular order imposed and therefore a tiny bit of energy. When the field is removed, thermal agitation returns the molecules to disorder, in the relaxation time t, and thermal energy is released. At 2450 MHz, the alignment of the molecules followed by their return to disorder occurs

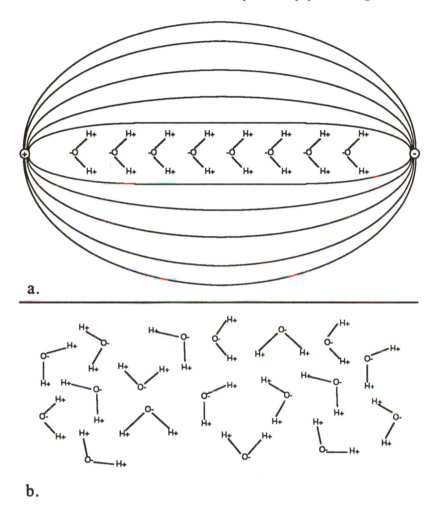

a.

b.

Figure 2.2. Schematic of the molecular response to an electromagnetic field. (a) polarized molecules aligned with the poles of the electromagnetic field; (b) thermally induced disorder as electromagnetic field is removed.

4.9×10^9 times per second, and results in very rapid heating. However, the efficacy of heating by dipole rotation depends upon the sample's characteristic dielectric relaxation time, that in turn depends upon temperature and the viscosity of the sample.

Effect of Dielectric Relaxation Time on Dipole Rotation. The dielectric relaxation time is the time that it takes for the molecules in the sample to achieve 63% of their return to disorder. The maximum energy conversion per cycle by many materials (dielectric loss due to dipole rotation) will occur when $\omega = 1/\tau$ where ω is the angular frequency of the microwave energy in radians per second ($\omega = 2\pi f$; f = microwave frequency) and τ is the dielectric relaxation time of the sample (9). A nonionic polar sample with a $1/\tau$ close to the angular frequency of the input microwave energy will have a high dissipation factor. In contrast, when $1/\tau$ of the sample is considerably different from the microwave angular frequency, the dissipation factor of the sample will be low.

Figure 2.3 illustrates the relationship between input microwave frequency and dielectric relaxation time on microwave penetration. The half-power depth for water is about 4 in. for 915 MHz and about 1 in. for 2450 MHz. The reciprocal of the dielectric relaxation time ($1/\tau$) for water is greater than 2450 MHz, so as the input frequency is reduced the difference between $1/\tau$ of water and the input microwave frequency increases. Therefore, absorption of the input energy is decreased and deeper penetration results.

As the sample is heated, the dielectric relaxation time will change as will the dissipation factor, and, therefore, the penetration depth. Table 2-2 illustrates this point. As the temperature of water is raised, the dissipation

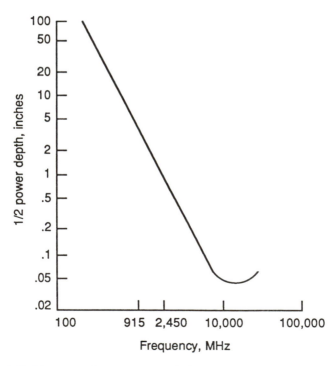

Figure 2.3. Variation of penetration with frequency for water at 25 °C. (Reproduced with permission from ref. 13. Copyright 1975 AVI Publishing).

Table 2-2. Effect of Temperature on the Dissipation
Factor of Water

Temperature (°C)	Tangent[a] ∂ ($\times$ 10)
1.5	3100
5.0	2750
15.0	2050
25.0	1570
35.0	1270
45.0	1060
55.0	890
65.0	765
75.0	660
85.0	547
95.0	470

[a]Measurements made at 3000 MHz. Data extracted from reference 9.

factor decreases. This decrease occurs because the $1/\tau$ of water increases, as the water temperature increases, and therefore the rotational frequency of the water is further out of coincidence with the input microwave angular frequency, and absorption decreases.

Effect of the Sample Viscosity on Dipole Rotation. A sample's viscosity affects its ability to absorb microwave energy (dissipation factor) because it affects molecular rotation. The effect of viscosity is best illustrated by considering ice water. When water is frozen, the water molecules become locked in a crystal lattice. This locking greatly restricts molecular mobility and makes it difficult for the molecules to align with the microwave field. Thus, the dielectric dissipation factor is low, 2.7×10^{-4} at 2450 MHz. When the temperature of the water is increased to 27 °C, the viscosity has decreased, and the dissipation factor is 12.2, which is much higher.

This situation should not be confused with the effect of the dielectric relaxation time. As just discussed, the reciprocal of the dielectric relaxation time ($1/\tau$) of water is greater than 2450 MHz; therefore, the increase in temperature from 0 to 27 °C should have decreased the dissipation factor. The viscosity of the ice water had a much greater impact on the dissipation factor than the dielectric relaxation time. For example, the dissipation factors for water at 45 and 95 °C are 7.5 and 2.4, respectively. Thus, as the water became more fluid (less viscous), the dielectric relaxation time had a greater impact on the dissipation factor of the water.

Relative Contributions of Dipole Rotation and Ionic Conduction

To a great extent, temperature determines the relative contributions of each of the two energy conversion mechanisms (dipole rotation or ionic conduction). For small molecules, such as water and other solvents, the dielectric loss to a sample due to the contribution of dipole rotation decreases as the sample temperature increases. In contrast, dielectric loss due to ionic conduction increases as the sample temperature increases. Therefore, as an ionic sample is heated by microwave energy, the dielectric loss to the sample is initially dominated by the contribution of dipole rotation. As the temperature increases, the dielectric loss is dominated by ionic conduction.

The percent contribution of these two mechanisms of heating depends upon the mobility and concentration of the sample ions and the relaxation time of the sample. If the ion mobility and concentration of the sample ions are low, then sample heating will be entirely dominated by dipole rotation. Therefore, as discussed in the section on dipole rotation, the heating time will then depend upon whether the reciprocal of the dielectric relaxation time $(1/\tau)$ is much higher or lower than the input microwave angular frequency. If $1/\tau$ is much higher or lower than the input microwave angular frequency, then heating time will be increased. On the other hand, as the mobility and concentration of the sample ions increases, microwave heating will be dominated by ionic conduction and the heating thime will be independent of the relaxation time of the solution. As the ionic concentration increases, the dissipation factor will increase and heating time will decrease (see Table 2-1). Heating time also depends on the microwave system design and the sample size and not uniquely on the dielectric absorptivity of the sample.

Sample Size

The input microwave frequency also affects the penetration depth of the microwave energy. As mentioned in the section on dielectric loss, the larger a sample's dissipation factor at a particular input frequency, the less it is penetrated by microwave energy. In large samples with high dissipation factors, the heating that occurs beyond the penetration depth of the microwave energy is due to thermal conductance through molecular collisions. Therefore, temperatures at or near the surface will be higher. Because boiling and other agitation increases the rate of thermal conductance, surface heating is not a problem unless penetration is very low and sample heating is extremely superficial. In that case, heat loss through the vessel walls can become significant and an increase in sample heating time will occur.

Although the small sample size used in most analytical dissolutions has advantages; it also presents at least one disadvantage. Figure 2.4 illustrates the decrease in the amount of absorbed microwave energy with a decrease in the sample size. (In this case the sample is water.) With small sample sizes, a considerable amount of energy is unabsorbed (reflected). Reflected energy can cause damage to the magnetron; therefore, in analytical work with small samples it is wise to use a microwave system that is designed to

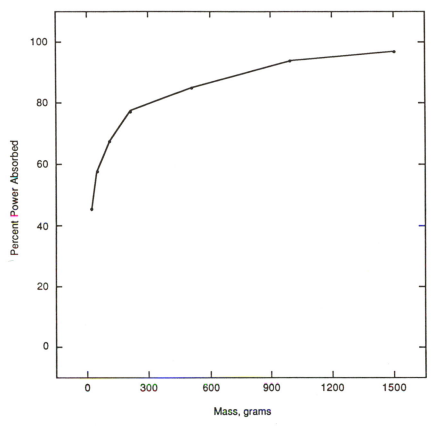

Figure 2.4. Percent microwave power absorbed by water. (Reproduced with permission from ref. 21.)

protect the magnetron from reflected power (*see* the section on the magnetron).

Microwave Heating

The typical time required to complete a wet digestion by conductive heating is from 1 to 2 h. In some instances it can be much longer. On the other hand, open-vessel dissolutions by microwave heating can be completed in 5–15 min (*12*). The difference is due to the sample heating method. Because vessels used in conductive heating are usually poor conductors of heat, it takes time to heat the vessel and transfer that heat to the solution. Also, because vaporization at the surface of the liquid occurs, a thermal gradient is established by convection currents, and only a small portion of the fluid is at the temperature of the heat applied to the outside of the vessel (Figure

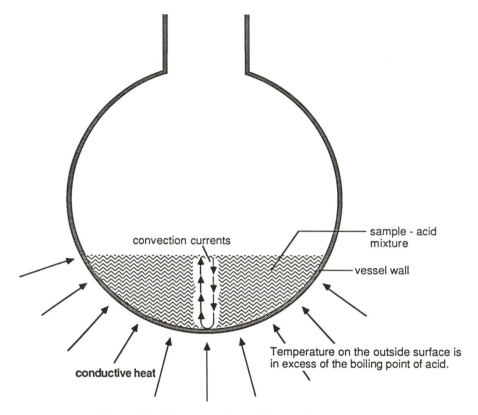

Figure 2.5. Schematic of sample heating by conduction.

2.5). Therefore, when conductively heating, only a small portion of the fluid is above the boiling point temperature of the solution.

On the other hand, microwaves heat all of the sample fluid simultaneously (for typical analytical sample sizes) without heating the vessel. Therefore, when microwave heating, the solution reaches its boiling point very rapidly. Because the rate of heating is so much more rapid, substantial localized superheating can occur (Figure 2.6).

Speed of Heating

Three predominant variables influence how rapidly a sample will be heated with microwave energy. These variables are the conductivity of the acid used, the dielectric relaxation time of the acid, and the acid volume. The dielectric relaxation time of the acid cannot be changed; however, the microwave frequency now available with analytical instruments is adequate to rapidly heat samples.

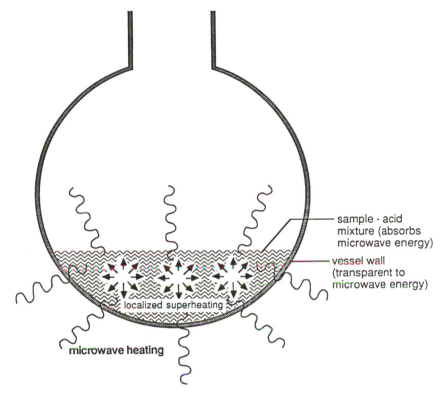

Figure 2.6. Schematic of sample heating by microwave energy.

Most analytical samples are small, and the sample is completely penetrated by the microwave energy. Because of this penetration, surface cooling problems are avoided, but the result is a greater risk of equipment damage from reflected energy. Current analytical microwave instruments protect the magnetron and thus reflected energy is less likely to affect the sample heating.

Microwave Instrumentation

The typical microwave instrument used for heating analytical samples consists of six major components: the microwave generator (called the magnetron), the wave guide, the microwave cavity, the mode stirrer, a circulator, and a turntable. Microwave energy is produced by the magnetron, propagated down the wave guide, and injected directly into the microwave cavity where the mode stirrer distributes the incoming energy in various directions (Figure 2.7). As discussed in the theory section, the percentage of the incoming energy that is absorbed depends upon the sample size and dissipation factor.

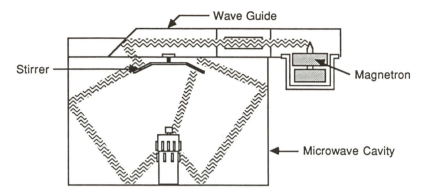

Figure 2.7. Schematic of the microwave cavity, wave guide, and magnetron.

The Magnetron

The magnetron is as a cylindrical diode with an anode and cathode. Superimposed on the diode is a magnetic field that is aligned with the cathode. A ring of mutually coupled resonant cavities is in the anode so that as a potential of several thousand volts is reached across the diode, the released electrons, under the influence of the magnetic field, resonate, and the magnetron oscillates (Figure 2.8). The oscillating electrons surrender energy to the microwave field that radiates from an antenna enclosed in the vacuum envelope of a tube (Figure 2.9).

In a fixed-tuned magnetron, the oscillations are designed to release the microwave energy at a certain frequency. Most microwave instruments used

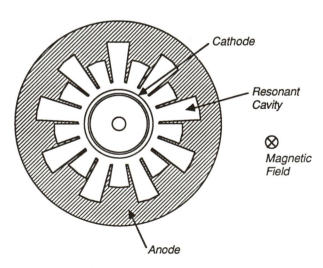

Figure 2.8. Schematic of the magnetron "diode".

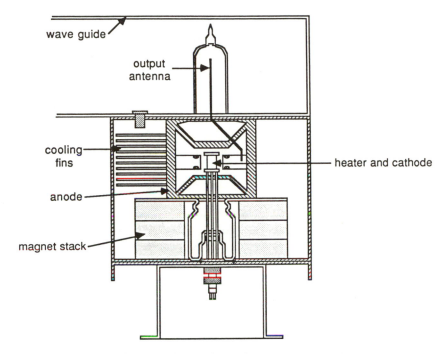

Figure 2.9. Schematic of a fixed-tuned magnetron (Hitachi 2M170).

for acid dissolutions have a fixed-tuned magnetron with an output frequency of 2450 ± 13 MHz. In these instruments, the magnetron receives approximately 1200 W of electrical line power that is converted to 600 W of electromagnetic energy. The remaining energy is converted to heat that must be dissipated by air cooling.

Cycling the Magnetron. In microwave systems used for sample preparation, the power output of the magnetron is controlled by "cycling" the magnetron to obtain an average power level. The duty cycle of a magnetron is the time the magnetron is on divided by the time base. For example, a time on of 5 s with a time base of 10 s is a duty cycle of 0.5. Likewise, a time on of 0.5 s with a time base of 1 s is a duty cycle of 0.5. The time base for the magnetron in microwave systems used for sample preparation is 1 second. Thus, to obtain one-half the rated output of 600 W the magnetron would be on for 0.5 s and off for 0.5 s. Appliance grade microwave ovens typically have a time base of 10 s. This relatively long time base used in the home microwave ovens is not desirable for heating analytical samples because heat losses can be significant during the long off time (5 s for a duty cycle of 0.5).

Power Output of the Magnetron. The microwave energy output from the magnetron is generally measured in watts (1 W = 14.33 cal/min), and is typically 600–700 W in microwave systems used for acid dissolutions. The power output of the magnetron can be indirectly determined by measuring the temperature rise of a quantity of water large enough to absorb essentially all of the energy delivered to the microwave cavity. Ordinarily, the apparent power output is determined by measuring the rise in temperature, in degrees centigrade, of 1 L of water heated at full power for 2 min (13). The general relationship used for evaluating the apparent power output is

$$P = C_p K \Delta T m / t \qquad (2.1)$$

where P is the apparent power absorbed by the sample (in watts); K is the conversion factor (from thermal chemical calories to watts); C_p is the heat capacity (or thermal capacity in calories per degree); $\Delta T = T_f\text{-}T_i$ (final temperature – initial temperature, in degrees centigrade); m is the mass of the sample (in grams); and t is time (in seconds). When water is used for performing an apparent power measurement, equation 2.1 can be simplified to

$$P = 35 \times \Delta T \qquad (2.2)$$

where 35 is the combination of the heat capacity of water, the conversion factor, time, and mass. The validity of the power test depends on a standard placement within the cavity and use of the same container. Because the dielectric dissipation factor and radiant losses are a function of temperature, the same initial temperature and approximate ΔT are used. The power test is more accurate if the starting temperature is 20 ± 2 °C each time.

Effect of Reflected Energy on Magnetron Function. The power output of a magnetron can be affected by overheating resulting from reflected microwaves. *Reflected power* occurs when the traveling electromagnetic waves are reflected and the flow of energy is partly in the reverse direction: toward the magnetron. Methods that transfer traveling waves from one medium to another without reflectance are called *impedance matching* (14). If the microwaves travel from the magnetron to the sample and are not reflected, then the system is perfectly matched. If there is reflection, then the system is said to be mismatched. Mismatching may cause overheating of the magnetron, a loss in output power, or even destruction of the magnetron.

A sample in the microwave cavity that has a high dissipation factor would be a model of a perfectly matched microwave system by absorbing nearly 100% of the input power. Microwave dissolutions usually involve the use of small volumes of acids that do not have extremely high dissipation factors at 2450 MHz. This situation results in reflected microwaves that in turn result in a mismatch between the magnetron and the microwave cavity

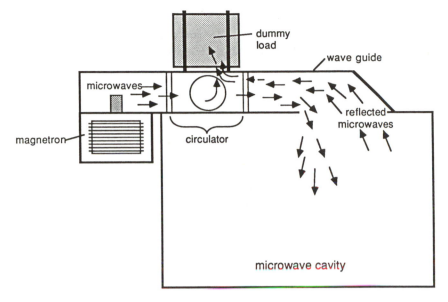

Figure 2.10. Schematic of a circulator deflecting the reflected microwaves to a dummy load.

and can change the output of the magnetron because of excessive heating of the magnetron.

Devices that remove reflected microwaves have been designed to protect the magnetron when mismatching occurs and to maintain a consistent power output. These devices are usually not available in appliance grade microwave ovens but are included in microwave systems used for acid dissolutions (*15*). The device most commonly used in these systems is a terminal circulator. The terminal circulator is a device that uses ferrites and static magnetic fields to allow microwaves to pass in the forward direction but diverts the reflected waves into a dummy load where the energy is harmlessly dissipated as heat (Figure 2.10).

The Wave Guide

The microwaves generated by the magnetron are channeled to the applicator (microwave cavity) by the wave guide. Wave guides are constructed of reflective material such as sheet metal, and are designed to direct microwaves to the cavity without a mismatch.

The Mode Stirrer

The mode stirrer is a fan-shaped blade that is used to reflect and mix the energy entering the microwave cavity from the wave guide. A mode stirrer assists in distributing the incoming energy so that the heating of the sample will be more independent of position. Appliance grade microwave ovens come with a predesigned mode stirrer that may be adequate for certain microwave dissolution applications.

The Microwave Cavity

The sample applicator into which microwaves are propagated is the microwave cavity. Simply stated, the microwaves entering the cavity are repeatedly reflected from wall to wall. The pathway of the microwaves is well-defined into recognizable patterns or modes that have have a beginning and end, as suggested in Figure 2.11. The direction and the number of modes excited in a cavity of a certain size are considered in the attainment of uniform microwave heating.

The microwaves entering the cavity intercept absorptive samples placed inside the microwave cavity, and lose energy with each interaction until no energy remains in a given wave. As discussed in the section on reflected energy and magnetron function, when a sample has a low dissipation factor, the microwaves continue to be reflected, and have a greater chance of finding their way back to the magnetron. Because the microwave cavity is constructed of metal, corrosion of the cavity is a concern for acid dissolutions in a microwave. Special modifications to protect the microwave cavity from corrosion are discussed in the following section.

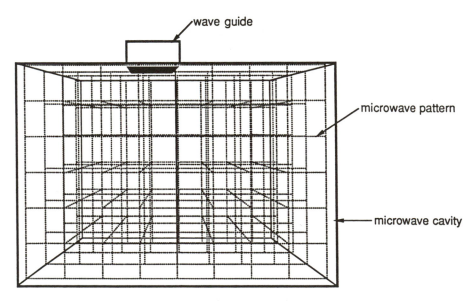

Figure 2.11. Schematic of a microwave pattern within a cavity.

Components and Accessories

As Figure 2.12 shows, there are three classes of microwave materials: reflective (metals), transparent (low-loss), and absorptive (high-loss). As just discussed, the water–acid solutions used in microwave dissolutions are absorptive materials, and the material used for construction of the primary components of microwave units are reflective materials. In this section, the components and accessories used inside the microwave cavity, most of which are constructed of transparent materials, will be discussed.

Attachment of Accessories to a Microwave System

In the early work with microwave heating, analytical chemists attempted to modify appliance grade microwave ovens in order to attach accessories such as monitoring equipment and scrubbers. For example, holes were drilled in microwave cavities to run tubes in and out to attach such accessories. If

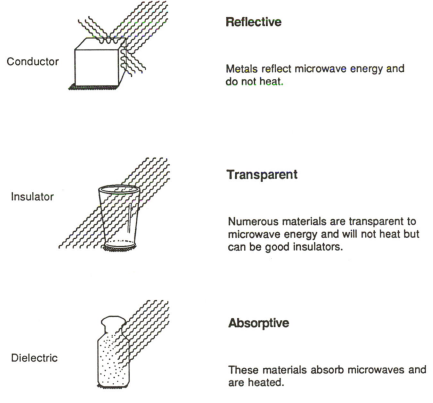

Reflective

Conductor

Metals reflect microwave energy and do not heat.

Transparent

Insulator

Numerous materials are transparent to microwave energy and will not heat but can be good insulators.

Absorptive

Dielectric

These materials absorb microwaves and are heated.

Figure 2.12. Interaction of materials with microwaves.

a user does not have the proper engineering background, modifications can lead to serious safety hazards such as microwave leakage. In many cases, an antenna effect occurs through such tubes, and a high percentage of the microwave energy entering the cavity is dumped externally. A user who is modifying an appliance grade microwave oven should be familiar with government regulations on microwave leakage and have the proper equipment to monitor the leakage (see Chapter 11). Analytical instruments are currently available that safely accommodate such accessories as temperature- and pressure-monitoring devices (15) and scrubbers (16).

It is theoretically possible to monitor any parameter while heating a sample in the microwave. The only limitation is that the monitoring probes must not perturb the microwave energy. In reality, this limitation is very big, because most probes will absorb microwave energy, behave as antennas, or arc in the microwave field. Monitoring systems are available for measuring temperature and pressure in a microwave system and are discussed in other chapters of this book.

Digestion Vessels

Containers should be constructed from low-loss materials so that the microwaves will not be absorbed by the vessel but will pass through the vessel to the solutions inside. Table 2-3 gives the dissipation factors of some common materials. From the standpoint of microwave transparency, Teflon [poly(tetrafluoroethylene)] and polystyrene and are excellent materials for construction of microwave accessories. Nylon, however, is not a good material. Fused quartz and glass are also good materials for use in a microwave

Table 2-3. Dissipation Factor of Different Materials

Material	Temperature (°C)	Tangent[a] ∂ (× 10)
Water	25	1570.0
Fused Quartz	25	0.6
Ceramic F–66	25	5.5
Porcelain No. 4462	25	11.0
Phosphate Glass	25	46.0
Borosilicate Glass	25	10.6
Corning Glass No. 0080	25	126.0
Plexiglass	27	57.0
Nylon 66	25	128.0
Polyvinyl Chloride	20	55.0
Polyethylene	25	3.1
Polystyrene	25	3.3
Teflon PFA	25	1.5

[a]Measurements made at 3000 MHz. Data extracted from reference 9.

environment. Other potentially good materials for use as vessels are poly-sulfone- and fiberglass-reinforced epoxy. Quartz, glass, and especially plastics are transparent to microwave energy and are poor conductors of heat; they are therefore good insulators when used as vessels for microwave heating.

The two basic considerations in choosing materials to construct vessels for use in acid dissolutions are chemical and thermal stability. Most of the plastics shown in Table 2-3 have good chemical resistance to acids but do not have good thermal stability at the temperatures required for acid dis-solutions. On the other hand, quartz and glass have good thermal stability but cannot be used for dissolutions that require hydrofluoric acid.

Teflon PFA is an ideal material for vessel construction for almost all acid dissolution applications because it is resistant to all acids and has a melting point of approximately 306 °C. Teflon PFA vessels can be used with conventional heating methods, but their use is restricted because Teflon PFA is an extremely poor conductor of heat. On the other hand, Teflon PFA vessels are ideal for use in microwave dissolutions. Because Teflon PFA is transparent to microwaves, the solution inside the vessel is heated directly and the vessel walls act as insulators. The only application in which Teflon could not be used for microwave dissolutions would be those dissolutions which use phosphoric acid or concentrated sulfuric acid. Both of these acids have boiling points above the melting point of Teflon PFA. As Table 2-3 indicates, quartz is an excellent material for use with either phosphoric or sulfuric acid in microwave dissolutions.

Open-Vessel Digestion.   In general, all vessels used for open-vessel digestions by conductive heating methods, except metal vessels, can be used in open-vessel digestions with microwave heating. A wide range of vessels such as glass beakers, boiling flasks, and refluxing vessels can be used. The development of vessels for specific applications in microwave dissolutions is limited only by the analytical chemist's imagination. It is preferable to construct the vessels of Teflon PFA because of its chemical and thermal stability.

In most open-vessel digestions, it is desirable to achieve as much re-fluxing as possible so that the need for continual additions of acid to maintain volume is decreased. Refluxing is generally better in open-vessel microwave dissolutions because the vessel is secondarily heated as heat from the solution is dissipated to the vessel and remains cooler. Therefore, with good air flow through the cavity, as is the case with analytical microwave systems, refluxing is better.

Closed-Vessel Digestion.   Closed-vessel systems for acid dissolutions have a number of advantages:

1. A closed-vessel digestion will achieve higher temperatures because the boiling point of the acid is raised by the pressure

produced in the vessel. With the higher temperatures produced in a closed vessel, the time required for digestion can
be greatly reduced.

2. The possibility of losing volatile elements during digestion is
   virtually eliminated in a closed vessel because there is little
   vapor loss.

3. Less acid is required. Because no evaporation occurs when
   digesting in a closed vessel, there is no need to continually
   add acid to maintain volume, one risk of contamination is
   eliminated.

4. The fumes produced by the digest are contained within the
   vessel, so there is no need to provide a method for handling
   potentially hazardous fumes.

5. With a closed-vessel digestion the possibility of airborne contamination is eliminated or substantially reduced.

Closed-vessel digestions can be done either by conventional heating methods
or by microwave heating. Closed-vessel digestion in a microwave, however,
requires far less time. With conventional closed-vessel methods, heat must
be conducted through the thick-walled material that is required to withstand
the increased pressure. Heat-up and cool-down times become excessive.
Therefore, conventional closed-vessel methods are used only for samples
that are very difficult to digest or for those samples in which the loss of
volatiles is a concern. Traditional closed-vessel methods are not practical
to use for most samples. However, in closed-vessel microwave digestions,
the solution is heated directly and the heat-up and cool-down times are
essentially the same as open-vessel digestion times. Thus, closed-vessel microwave digestions have all the advantages of open-vessel digestions and
dramatically reduce the time required to digest nearly all sample types (17).

An example of a closed vessel used for microwave digestions is given
in Figure 2.13 (18). This vessel consists of a vessel body, vessel cap, and
safety relief valve. The vessel is made of Teflon PFA and designed to operate
at internal pressures up to 120 ± 10 psi. At this pressure the safety valve
will open and then reseal.

Turntable System

As mentioned in the section on the microwave cavity, when microwaves
enter the cavity, they are distributed in patterns (modes) by reflection from
the metallic surfaces of the cavity. When a single vessel is placed inside the
cavity, the microwave energy intercepts the vessel from all directions. As
the waves intercept the vessel, a percentage of the energy in the wave is

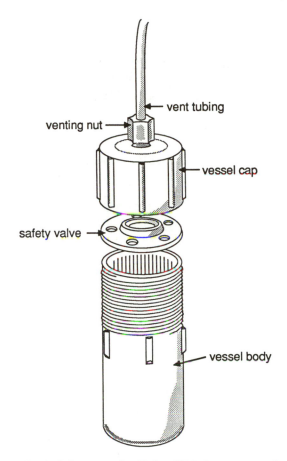

Figure 2.13. Schematic of a Teflon PFA digestion vessel.

absorbed according to the dissipation factor of the solution in the vessel. However, if two vessels are placed in a cavity, the vessel location within the cavity determines the level of exposure of the vessels to the established microwave pattern (Figure 2.14). For example, if two vessels are placed in a microwave cavity, one vessel might be in a position of reinforcement of the criss-crossing full-cavity-width waves, and the other in a position of partial cancellation. If one vessel has greater exposure to the established microwave pattern than the other, it will heat differently, and nonuniformity of heating will occur.

The uniformity of heating multiple vessels can be greatly increased by rotating the vessels on a turntable (Figure 2.15). Turntables are available that rotate 360° continuously, or that alternate back and forth 180°. The alternating rotation is used when monitoring devices, such as a pressure monitor, are connected to the reaction vessels.

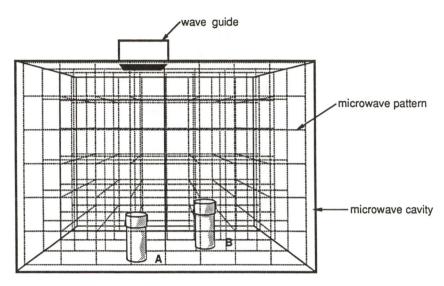

Figure 2.14. Schematic of a microwave pattern interaction with digestion vessels.

Microwave Furnaces

Conventional furnaces are used to dissolve samples that are very difficult to digest. They are used for performing fusions and for dry ashing samples with high percentages of organic material before acid dissolution. Microwave furnaces have also been developed (19) for these applications (20). The microwave furnace uses a material with a very high dissipation factor, such as silicon carbide, to absorb essentially 100% of the input power in a small area surrounded by a quartz insulation (Figures 2.16 and 2.17). With a 10,000-cal input/min, the furnace can achieve a temperature of 1000 °C in approximately 2 min. The advantage of the microwave furnace relative to the conventional furnace is that the conventional furnace must be kept at temperature at all times, because it takes so much of time to bring it up to temperature. A considerable amount of energy is consumed by operating the furnace continuously. There are no heating coils to burn out in the microwave system as in the conventional furnaces. Another advantage of the microwave furnace is that the user is not exposed to the heat as the sample is added or removed.

Summary

The purpose of this chapter was twofold. First, it was intended to give the analytical chemist a theoretical understanding of microwave heating and its relation to sample preparation. Second, it described how the microwave

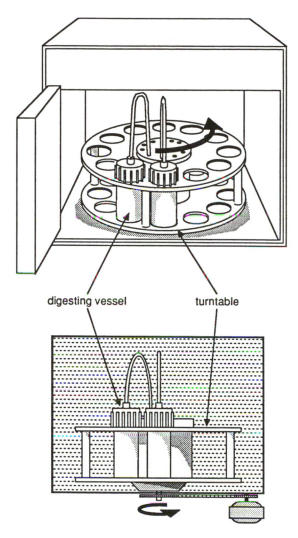

digesting vessel   **turntable**

Figure 2.15. Schematic of a turntable system used for rotating digestion vessels in the microwave cavity.

hardware, accessories, and components function in analytical sample prep-aration. Microwave heating is quite different from traditional conductive heating methods, but microwave heating has some unique advantages, per-haps the most important of which is speed. Also, the necessary microwave hardware and various tools are available now to allow the analytical chemist to benefit from the advantages that microwave heating offers.

Most of what has been discussed in this chapter and is discussed in this book involves the use of microwave heating with strong acids to accomplish

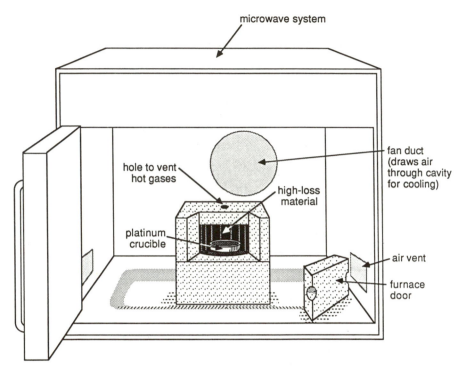

Figure 2.16. Schematic of a microwave muffle furnace.

wet digestions. This application for microwave heating has just begun, and should expand dramatically in the next few years. Microwave heating should be used in many areas of analytical chemistry in the future. Some natural extensions of the wet digestion application include protein and carbohydrate hydrolysis for monomer determination and solvent extractions. Sophisticated analytical instruments are currently used in these applications for very short analysis times following hydrolysis or extraction, but the heating procedures for both of these applications are long and tedious.

As with any emerging technology, new applications will lead to new developments and improvements in microwave hardware. One area of hardware development that has already begun is the use of microwave heating with robotics (*see* Chapters 9 and 10). In the future there may be automated microwave hardware that does not require the use of robotics. Perhaps another hardware development will be a complete flow injection system in which the sample will be be digested as it flows through a microwave heating chamber on its way to being analyzed. As mentioned in the section on open-vessel digestion, the possibility for new vessel designs for microwave heating applications is unlimited.

The use of microwave energy for solving long-term heating problems

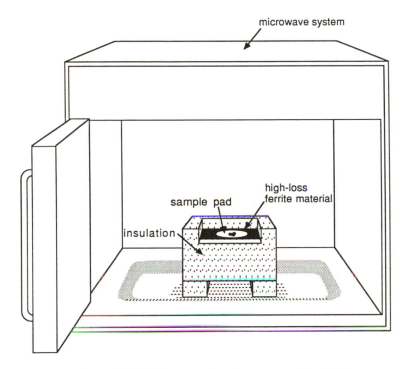

Figure 2.17. Schematic of a microwave ashing block.

in the analytical laboratory is in its infancy. There should be many exciting developments in the years to come as the theoretical concepts of microwave interaction with analytical samples is better understood and hardware capabilities increase.

Literature Cited

1. Spencer, P. L. U.S. Patent 2 605 383, 1952.
2. Bluhm, D. D.; Markuszewski, R.; Richardson, C. K.; Fanslow, G. E.; Durham, K. S.; Green, T. Abstracts of Papers. British National Committee for Electroheat, Heating, and Processing Conference. Cambridge, UK: Energy Research Abstracts, 1986; Abstract ERA8608.
3. Smith, R. D. *Elec. Power Res. Inst. EM-3645,* **1984,** 2-1.
4. Bostian, M. L.; Fish, D. L.; Webb, N. B.; Arey, J. J. *J. Assoc. Off. Anal. Chem.* **1985,** 68, 876.
5. Collins, M. J.; Cruse, B. W.; Goetchius, R. J. U.S. Patent 3 909 598, 1975.
6. Collins, M. J.; Cruse, B. W.; Goetchius, R. J. U.S. Patent 4 457 632, 1984.
7. Collins, M. J. U.S. Patent 4 554 132, 1985.

8. Mudgett, R. E. *Food Tech.* **1986,** June.
9. Von Hippel, A. R. *Dielectric Materials and Applications*; John Wiley: New York, 1954; p 301.
10. Smith, R. D. Elec. Power Res. Inst. EM–3645, 1984, p A-8.
11. Decareau, R. V. *Microwaves in the Food Processing Industry*; Academic: New York, 1985. Chap 1.
12. Nadkarni, R. A. *Anal. Chem.* **1984,** 56, 2233.
13. Copson, D. A. *Microwave Heating*; AVI: Westport, CT, 1975; Chap 1.
14. Crawford, F. S. *Waves*; McGraw–Hill: New York, 1968; Chap 4.
15. Kingston, H. M.; Jassie, L. B. *Anal. Chem.* **1986,** 58, 2534.
16. White, R. T.; Douthit, G. E. *J. Assoc. Off. Anal. Chem.* **1985**, 68, 766.
17. Fisher, L. B. *Anal. Chem.* **1986**, 58, 261.
18. Revesz, R.; Hasty, E. T. Presented at the Pittsburgh Conference, Atlantic City, NJ. March 1987; paper 252.
19. Collins, M. J.; Hargett, W. P. U.S. Patent 4 565 669, 1986.
20. Matthes, S. Presented at the Eastern Analytical Symposia, New York, October 1986; paper 072.
21. Kingston, H. M.; Jassie, L. B. Presented at the Eastern Analytical Symposia, New York, October 1986: paper 076.

RECEIVED for review April 10, 1987. ACCEPTED revised manuscript April 4, 1988.

CHAPTER 3

Guidelines for Developing Microwave Dissolution Methods for Geological and Metallurgical Samples

S. A. Matthes

"Basic research is when I'm doing what I don't know I'm doing".
Werner von Braun

Guidelines for developing microwave dissolution methods for geological and metallurgical samples are being investigated at the Bureau of Mines. By varying sample size, acid mixture, vessel type (open or closed), microwave power, and digestion times, procedures have been developed to successfully dissolve metals and minerals. A general scheme of microwave digestion and fusion methods is presented, as well as specific techniques for the dissolution of phosphate rock, ores, solder, superalloys, and other geological and metallurgical samples. Development of specific methods involves the selection of parameters that best exploit the chemistry of the sample matrix and the dissolution-enhancing effects of microwave radiation.

ANALYTICAL RESEARCH AT THE BUREAU OF MINES is aimed at improving the methods used to characterize geological and metallurgical samples. An important aspect of this research is the reduction in time, complexity, and expense of sample dissolution for chemical and instrumental analysis. Rapid, low-cost methods for the acid, pressure dissolution of slags, which are in sealed plastic bottles heated in either a pressure cooker (*1*) or a boiling water bath (*2, 3*) have been studied at the Bureau of Mines. Because the bottles used in these methods were transparent to microwave radiation, a modified kitchen-type microwave appliance was used in place of the water bath to heat the samples (*4*). Later research has extended the use of microwave methods to include dissolution of glasses, cements, superalloys and difficult geological samples (*5, 6*).

Figure 3.1. CEM microwave dissolution system including the MDS-81D microwave unit, Teflon digestion vessels, Teflon turntable, and capping station.

Other reports of microwave dissolution methods for geological and metallurgical samples were given by Nadkarni (coal, fly ash, oil shales, sediments, rocks) (7); Smith et al. (sulfide minerals) (8); Labrecque (sulfides, slags, ores) (9); Fischer (geological samples) (10); Fernando et al. (steels) (11); and Lamothe et al. (geological samples) (12). The introduction of the first commercial microwave dissolution system (Figure 3.1), having a Teflon [poly(tetrafluoroethylene)] dissolution vessel with a pressure-relief valve (Figure 3.2), has expanded the number of sample types amenable to microwave dissolution and provided the level of safety necessary for the adoption of microwave methods for general laboratory use.

The interest generated by the publication of the Bureau's first microwave dissolution method (4) has resulted in hundreds of requests for new microwave dissolution methods. On the basis of research stimulated by this interest, this laboratory has developed a general scheme of microwave techniques for dissolving geological and metallurgical samples. The purpose of this scheme is to provide the analyst with guidelines for developing microwave dissolution methods for these sample types.

Because of the varying chemistries of geological and metallurgical materials, a variety of procedures are required for the digestion of different sample types. Development of the specific methods involves the selection of parameters that best exploit the chemistry of the sample matrix and the

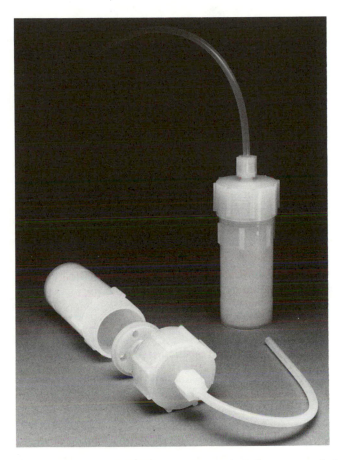

Figure 3.2. CEM PFA Teflon dissolution vessels with pressure-relief valve.

dissolution-enhancing effects of microwave radiation. By selecting the appropriate digestion vessel, acid mixture, microwave power, and digestion times, successful methods for the dissolution of most metals and minerals can be developed.

Microwave dissolution is particularly suited for rapid preparation of samples for instrumental analysis. Its use as a means of preparing analytical solutions for atomic absorption (AA) (4–6, 8, 9, 13) and inductively coupled plasma (ICP) emission spectrometry (7, 11, 12, 14) is well-documented. If complete dissolution is achieved, accuracy and reproducibility are generally at least as good as with conventional techniques, and microwave dissolution offers the added advantages of greater speed, lower cost, increased retention of volatiles, and reduced contamination. The aim of this study is not the analysis of solutions, but rather the preparation of analytical solutions by microwave dissolution. Solutions prepared by the following procedures are suitable for analysis by a wide range of instrumental and chemical methods and have been thoroughly tested to routinely yield good analytical results.

Microwave Conditions for Metallic and Mineral Samples

General Conditions for Microwave Dissolution

All work was performed in an MDS-81D microwave unit (CEM). PFA Teflon digestion vessels (CEM) fitted with relief valves were used, unless otherwise indicated. The vessels rotate on a Teflon turntable during the digestion to ensure uniform exposure of each sample to the microwave radiation. Caps were tightened at a capping station (CEM) to provide the uniform tightness necessary for the relief valves in each vessel to vent at the same pressure.

Samples should be in powder form, preferably finer than 100 mesh; however, metals may be turnings, filings, shavings, or fine wire. Concentrated reagent grade acids are usually of sufficient purity for most analyses, but ultrapure acids are recommended for trace element determinations.

Metallurgical Samples—Dissolution Considerations

Metallurgical samples interact with microwave radiation in many ways. Many metals and metallurgical samples present no difficulties, and dissolve readily and safely by microwave methods. However, several metal types, particularly Fe-based alloys, spark when subjected to microwave radiation. This sparking can present a serious safety hazard, particularly if dissolution is performed in a closed vessel. The dissolution of metals by acids forms H_2 gas, which can be ignited by microwave-induced sparking (11). When first attempting the dissolution of an untried sample, the sample should be allowed to react thoroughly with the acid before the vessel is placed in the microwave oven. If possible, the microwave digestion should be carried out in an open vessel. If a closed vessel is required to achieve dissolution, all O_2 should be purged from the vessel before sealing. Argon or nitrogen can be used to fill the vessel before capping, or the addition of acid can be made in a N_2-filled glove box.

The sample should be exposed to microwaves for the minimum required time. For many steel samples, a multiple additions technique is recommended. After allowing the sample to initially react with the acid for 2 min, the sample is digested in an open vessel in the microwave oven for 1 min followed by a second addition of acid and return of the sample to the microwave oven for an additional minute. This procedure can be repeated as required to achieve dissolution. Tall glass beakers and acid mixtures containing H_2SO_4 are very useful for this technique. Hot, concentrated H_2SO_4, H_3PO_4, and other reagents that have boiling points exceeding the maximum use temperature of Teflon are not suitable for use in Teflon vessels.

Microwave Dissolution Methods for Metallurgical Samples

Microwave digestion can be a particularly effective means to accelerate dissolution of metallic samples. Microwaves apparently prevent passivation, whereby the dissolution may cease because of the formation of an acid-resistant coating on the sample surface. Using mixtures of HF, HCl, and HNO_3 at elevated temperatures and pressures in closed vessels without loss of the HF is significant, because HF can suppress passivation. The general procedure for microwave acid pressure dissolution is summarized in Figure 3.3. For metal and some inorganic samples, a high density of microwaves

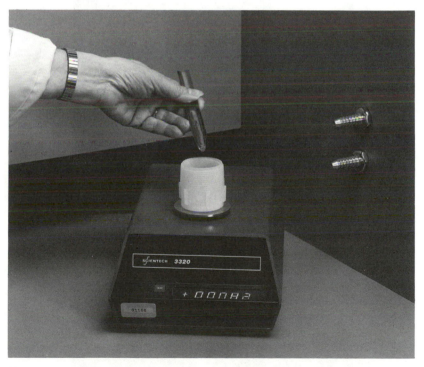

Figure 3.3. General procedure for acid, pressure dissolution in sealed Teflon vessels in the CEM microwave system. Step 1, sample is weighed into digestion vessel. Continued on next page.

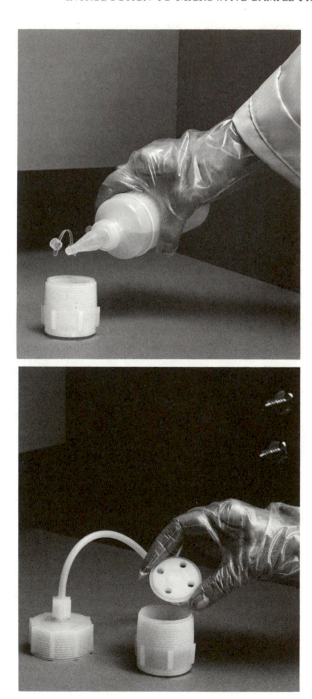

Figure 3.3. Continued. Top: Step 2, a suitable acid mixture is added. Bottom: Step 3, pressure relief valve is put in place.

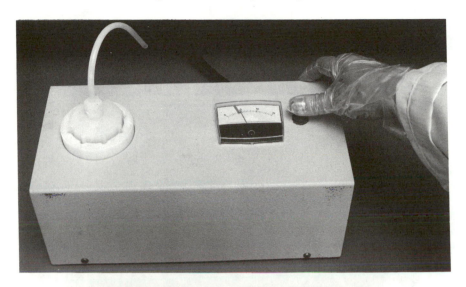

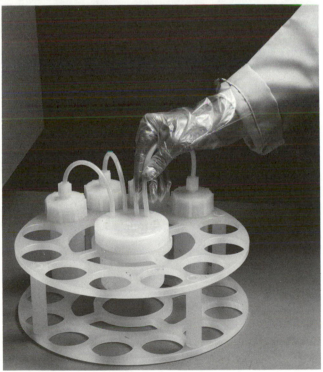

Figure 3.3. Continued. Top: Step 4, capping station provides a uniform tightness to the vessel screw caps. Bottom: Step 5, vessels are placed on the turntable with vent tubing connected to central catch vessel. Continued on next page.

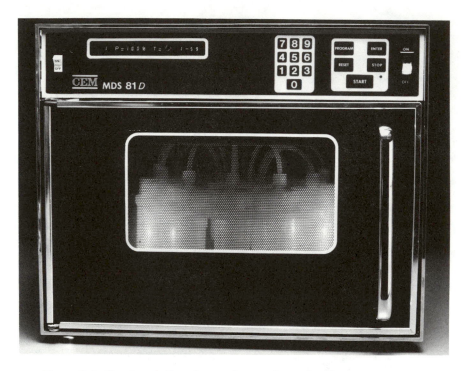

Figure 3.3. Continued. Step 6, samples are digested in the microwave unit.

probably exists at the sample–acid interface. This increased density of microwaves at the interface may be caused by reflection of unabsorbed microwaves. Acid molecules adjacent to the surface exposed to this higher density radiation are rapidly heated; heating creates turbulence that constantly sweeps clean the reaction surface.

The following microwave dissolution methods illustrate a general scheme for microwave digestion of metallurgical samples (Figure 3.4). Concentrated reagent grade acids are used in all methods.

Steel (NBS No. 121C)

Steels are ferrous-based alloys containing varying amounts of C, Ni, Cr, Mn, and other elements. Because of the great variety of steels, no one procedure can be used for all types. NBS No. 121C (Steel) is just one example:

1. Weigh 0.2-g sample into a 120-mL Teflon vessel.
2. Add 6 mL of HCl and 2 mL of HNO_3.
3. Let stand for 2 min, shaking occasionally.

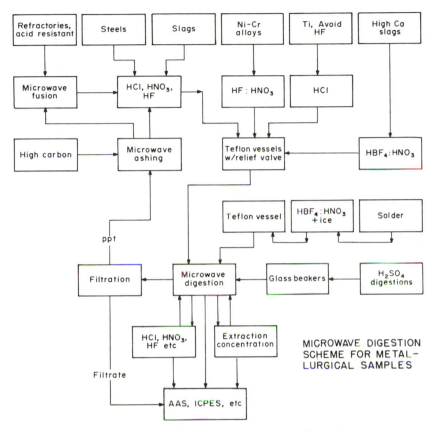

Figure 3.4. Microwave digestion scheme for metallurgical samples.

4. Place a relief valve on the vessel **Upside Down** and cap, so that the vessel is not sealed, but the splashing will be contained in the vessel.

5. Digest in microwave oven for 30 s at 50% power.

6. Wait 1 min, remove from the microwave, and add more acid as in Step 2.

7. Repeat Steps 3–5.

8. Remove from the microwave, add 1 mL of HF, and dilute to 50 mL with deionized H_2O.

Copper Reverberatory Slag

Slags are byproducts of the commercial production of metals from ore, and consist chiefly of SiO_2, CaO, Al_2O, Fe_2O_3, and a large number of other minor and trace constituents. Slags, glasses, and other amorphous silicate materials provide some of the best examples of the effectiveness of microwave–acid–pressure dissolution because they dissolve easily without loss of volatile SiF_4. The HF content of the acid mixture can be altered for varying concentrations of Si in the sample. For example, 10 mL of the HCl–HF mixture would be used for a 0.2-g borosilicate glass sample. Boric acid is conveniently added to the sample solutions to complex unreacted HF, to redissolve precipitated fluorides, to allow the use of conventional glassware, and to act as a matrix modififer for atomic absorption analysis (1–4).

1. Weigh a 0.1- to 0.2-g sample into a 120-mL Teflon vessel.
2. Wet sample with 2 mL of HNO_3 followed by 5 mL of a 7:3 HCl:HF mixture.
3. Seal with a cap and relief valve using a capping station.
4. Digest at 100% power for 2 min.
5. Cool. Remove the cap and dilute to 100 mL with 1.5% boric acid solution. If necessary, recap vessel, and return to microwave for an additional 2–8 min at 100% power to dissolve fluorides.

Superalloys

Heat- and corrosion-resistant Ni-, Cr-, and Co-based alloys (Inconels, Hastelloys, NBS No. 168, etc.) may require several days for dissolution by conventional methods, but are easily dissolved in a few minutes with microwaves. Metals generate large volumes of gas during acid digestion, and appropriate care should be taken following any dissolution in a sealed vessel to vent these gases. Cooling in an ice bath followed by careful opening of the vessel in a fume hood are strongly recommended.

1. Weigh a 0.1- to 0.5-g sample into a 120-mL Teflon vessel.
2. Add 10 mL of a 1:1:1 $HF:HNO_3:H_2O$ mixture.
3. Seal with a cap and relief valve using capping station.
4. Digest for 5 min at 100% power.
5. Cool. Remove cap and dilute to volume with deionized H_2O.

Pb–Sn Solder (NBS No. 127)

Some materials may dissolve easily, but contain constituents that have radically different solubilities. An example of this type sample is Pb–Sn solder. Tin, if dissolved in the presence of oxidizing acids and heat, will precipitate as metastannic acid; Pb precipitates as PbF_2 with the addition of HF. In this example, HBF_4 is used to prevent precipitation of PbF_2, and the temperature is reduced by adding an ice cube to the acid mixture. This procedure has the desirable effect of rapidly dissolving the metal at a hot sample–acid interface while maintaining a cool bulk-solution temperature. As a solid, ice will not absorb microwaves.

1. Weigh a 1-g sample into a 120-mL Teflon vessel.
2. Add 18 mL of HBF_4, 10 mL of H_2O, 4 mL of HNO_3, and one ice cube (approx. 1 in.³). Do not cap the vessel.
3. Digest at 100% power until the ice melts.
4. Dilute to volume with deionized H_2O.

TiO_2 Waste

TiO_2 waste contains 50% C and 50% TiO_2 with traces of Nb, Ta, Th, U, Fe, and V. To prevent precipitation of U and Th fluorides, HF is avoided. The high C content can also interfere with the recovery of the metals. The dissolution of this sample requires a preliminary ashing step to remove C followed by fusion of the resulting ash with Na_2O_2. A recently developed microwave muffle furnace (Figure 3.5) allows the ashing and fusion steps to be carried out in the microwave. The muffle furnace consists of a block of microwave-transparent firebrick containing a cavity large enough for one or two small crucibles. A SiC lining in the cavity absorbs microwaves very efficiently and will quickly reach temperatures that either "burn-off" any free C in an ashing step or melt common fluxing agents such as Na_2O_2, dissolving the acid-resistant samples in the molten flux. Fusion products produced in this manner are easily dissolved in mineral acids. Analytical solutions produced from fusions are frequently contaminated with Ca, K, and other elements, and require careful matching of the flux content in the standard and sample solutions. For this reason, and because of the high salt content of fusion solutions, fusions should never be used to achieve dissolution if an acid dissolution will do the job.

1. Weigh a 0.5- to 1.0-g sample into a Zr crucible. Place the crucible in cold microwave muffle furnace and heat at 100% power for 18 min
2. Cool. Add 3 to 4 g Na_2O_2 and heat in microwave muffle furnace for 16 min.
3. Cool. Acidify with 5% H_2SO_4 in a 400-mL beaker. Heat in a microwave oven until clear.

Figure 3.5. CEM microwave muffle furnace. The SiC-lined cavity of the furnace is an efficient microwave absorber and can reach ashing or fusion temperatures in a few minutes.

Microwave Muffle Furnace vs. Conventional Muffle Furnace

In the last example, a novel microwave-heated muffle furnace was used to quickly fuse a highly refractory sample in Na_2O_2. By using a microwave oven, a sample is sequentially ashed, fused, and brought into solution. By performing these tasks in a single appliance, tremendous savings are achieved in energy costs, laboratory space, and equipment. For relatively simple procedures involving a single sample and not requiring a high degree of temperature control, the microwave muffle furnace is very useful. However, because of the present small size of the cavity, and the lack of temperature feedback control, the microwave muffle furnace is inferior to the conventional muffle furnace, which can handle many samples with a much higher degree of temperature control. Future improvements in technology may improve the microwave muffle furnace, and allow its use in place of the conventional muffle furnace in a wider variety of methods.

Microwave Digestion Methods for Geological Samples

The dissolution of geological samples by microwave methods is complicated by the great variety of sample types. Many minerals (chromite, ilmenite, quartz, etc.) are extremely acid-resistant and defy the most vigorous attempts to dissolve them, even in the microwave. At present, the best way to attack these samples is by Na_2O_2 or $Li_2B_4O_7$ fusion. Highly refractory materials, such as chromite, require the more rigorous fusion with the very caustic molten Na_2O_2. Quartz and similar minerals can be fused easily by the less rigorous, higher temperature melt, $Li_2B_4O_7$. The development of the microwave muffle furnace makes these procedures possible in the microwave. Frequently, an acid-leaching procedure can successfully dissolve the elements of interest and leave behind the acid-resistant matrix. The general scheme for microwave digestion of geological samples in shown in Figure 3.6.

Phosphate Rock (NBS No. 120b)

Phosphate rock contains high levels of Ca and Si, so the use of HF to dissolve the Si may precipitate CaF_2. If Ca content is very high ($>50\%$), the CaF_2 may not redissolve after adding boric acid. Precipitation of CaF_2 can be avoided by using HBF_4 instead of HF. Portland cement (e.g. NBS No. 636) and high-Ca slags can also be prepared by this method.

1. Weigh a 0.1- to 0.5-g sample into a 120-mL Teflon vessel.
2. Add 10 mL of 1:1 HBF_4:HNO_3.
3. Seal with a cap and relief valve using a capping station.
4. Digest in a microwave unit at 100% power for 2 min.
5. Cool. Dilute to volume with deionized H_2O.

Alaskan Tin Ore

Quartz is very resistant to acid attack, even by HF. Many ores contain 80% quartz or more. Frequently, microwave digestion can leach the metallic fraction of the ore from the quartz, but in Alaskan tin ore the Sn was resistant to leaching by HF alone and HNO_3 was avoided to prevent the Sn from hydrolyzing. A $Li_2B_4O_7$ fusion method using the microwave muffle furnace was employed (Figure 3.7).

1. Weigh a 1-g sample into a Pt crucible.
2. Add 4 g of $Li_2B_4O_7$, mixing well.
3. Place into a cold microwave muffle furnace and heat for 24 min at 100% power.

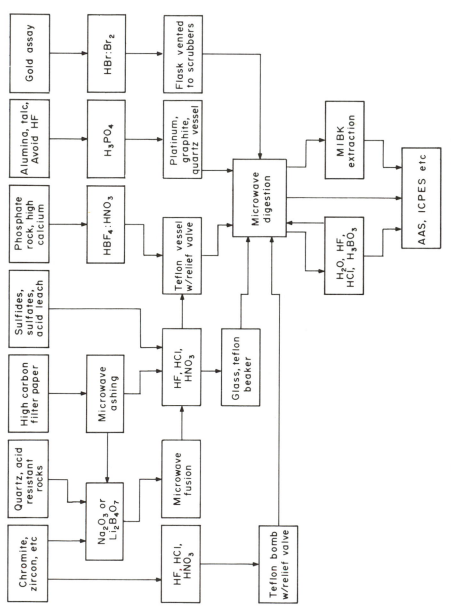

Figure 3.6. Microwave digestion scheme for geological samples.

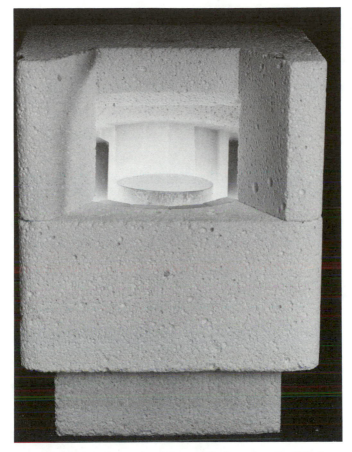

Figure 3.7. A $Li_2B_4O_7$ fusion in a Pt crucible heated in the CEM microwave muffle furnace.

4. Cool. Place fusion button into a 120-mL Teflon vessel.
5. Add 10 mL of HF and 90 mL of H_2O.
6. Tighten cap with relief valve using a capping station.
7. Digest at 100% power for 6 to 8 min.
8. Transfer to a 250-mL flask and dilute to volume with deionized H_2O.

Sulfide Ore

This method is used to dissolve sulfide minerals from mines, mills, and smelters. Samples consist of sulfides of Cu, Ni, and Fe with varying amounts of silicate minerals associated with the concentrated ore. This method completely dissolves the S, which is necessary for total recovery of Se and As for analysis. Dissolution in a sealed vessel with HNO_3 prevents the loss of volatile chlorides of these elements (9, 14).

1. Weigh a 0.3-g sample into a 120-mL Teflon vessel.
2. Add 3 mL of HNO_3.
3. Seal with a cap and relief valve with a capping station.
4. Digest in a microwave oven for 3 min at 40% power.
5. Cool. Add 5 mL of HCl and reseal with a cap and relief valve.
6. Digest 3 min at 40% power in a microwave.
7. Cool. Dilute to 30 mL with deionized H_2O.
8. Return to the microwave at 100% power until solutions boil.

Gold Ore

Gold assay requires a large sample size (>25 g). The Au may be present in the ore as metal, as alloy, or as a compound that makes simple acid digestion unsuitable. An HBr–Br_2 extraction followed by methyl isobutyl ketone (MIBK) extraction from the HBr solution is used in this method. For some sulfide ores (see Sulfide Ore), a preliminary aqua regia treatment may be necessary for complete Au recovery (13).

1. Weigh a 25 to 100 g of pulverized ore into a 250-mL flask.
2. Add HBr–Br_2 to the flask. (Note: Amount of HBr-Br_2 varies with the sample size; approximately 2 mL of acid per gram of sample is sufficient.)
3. Digest in a microwave unit at 50% power for approximately 20 min. The time will vary with sample size. (Note: Flask must be vented through a scrubber to a fume hood.)
4. Cool. Extract Au with MIBK.

Nickel, Copper Ores, Feeds, Concentrates, and Tails

These samples represent the various stages in the refining process of Ni–Cu ores similar to those in the section on sulfide ore. To maximize the number of samples processed for analysis, the loss of volatile species is ignored and dissolution is carried out in open glass test tubes in a special turntable that allows 40 simultaneous digestions in the microwave. This is a quick digestion procedure for the analysis of Ni, Cu, Co, and Fe in a large number of samples with varying composition (*14*).

1. Weigh a 0.25-g sample into a glass test tube.
2. Add 5 mL of HCl and HNO_3, 2 mL of H_2SO_4, and 1 mL of HF.
3. Place on a custom-made Teflon turntable (40-tube capacity) and digest for 10 min at 70% power.
4. Dilute with deionized H_2O as necessary.

Lead or Silver Bead from Fire Assay of Exploration Samples

The traditional means of determining the precious metals, Pt, Pd, Au, and Ag is by the fire-assay technique, in which the metals of interest are concentrated from a large sample of ore into a small metal bead. The following method is for the rapid determination of Pt, Pd, and Au in a Ag bead or Ag in a Pb bead to 0.002 oz/ton for a large number of bead samples (*14*).

1. Place the bead in test tube (Ag bead—10 mg, Pb bead—300 mg).
2. Add 3:1 HCl:HNO_3 for a Ag bead or 6:1 H_2O:HNO_3 for a Pb bead.
3. Place on custom-made Teflon turntable (40-tube capacity) and digest in a microwave for 10 min at 30% power.
4. Dilute as necessary with deionized H_2O.

Chromite

Chromite is acid-resistant and refractory, requiring a Na_2O_2 fusion method. The following method is also useful for dissolving chrome refractory (NBS No. 103a).

1. Weigh a 0.5-g sample into a Zr crucible.
2. Add 6 g Na_2O_2. Mix thoroughly.
3. Place in a cold microwave muffle furnace and heat at 100% power for 18 min. (Note: Zr will oxidize if heating time at 100% power exceeds recommendation.)
4. Remove crucible and cool.
5. Acidify and dissolve the fusion product with HCl in a 250-mL beaker.
6. Dilute to 250 mL with deionized H_2O in a volumetric flask.

Discussion

A material may either be reflective, absorbent, or transparent to microwaves. Many microwave transparent materials are useful as vessels (e.g., Teflon digestion vessels or glass beakers), allowing microwave radiation to pass through them to couple with an internal absorbing medium. Absorbing materials may be H_2O or acid molecules that are very quickly heated by their rapid alignment to the microwave radiation. Direct heating of the dissolution medium instead of heating by convection or conduction accelerates the speed of digestion. The solid sample itself may also absorb microwaves, creating a heated surface on which the acid can react. For example, SiC is an extremely efficient microwave absorber and is used as the lining of the microwave muffle furnace to rapidly reach ashing or fusion temperatures. Samples that reflect microwaves may have a high density of microwave radiation at the sample–acid interface, resulting in extremely rapid heating of the acid and in turbulence that sweeps the sample surface clean exposing fresh sample surface for dissolution. The sparking of metallic samples can present a safety hazard when metals react with acids because of the production of H_2.

Although no one procedure exists for the dissolution of all sample types, only a few different methods are required. These methods can be varied to suit the peculiarities of each material. Microwave ashing, fusions, Teflon or glass vessels, and closed-vessel or open- vessel digestion are parameters that may be combined with different acid mixtures, digestion times, and power settings to develop procedures that will dissolve almost any sample. The microwave digestion scheme presented here is the current state-of-the-art in microwave dissolution technology. Samples that require fusions for dissolution today may be successfully dissolved by closed-vessel acid digestion tomorrow. Improvements in vessel design, allowing digestions to take place at higher pressures, and new vessel materials that will enable the use of higher boiling acid, will add hundreds of sample types to the list of materials dissolved by microwave acid pressure dissolution.

Disclaimer

Reference to specific equipment, trade names, or manufacturers is made for identification only and does not imply endorsement by the Bureau of Mines.

Literature Cited

1. Farrell, R. F.; Mackie, A. J.; Lessick, W. R. *Rep. Invest.*— *U. S. Bur. Mines* **1979**, *RI* 8336.

2. Farrell, R. F.; Matthes, S. A.; Mackie, A. J. *Rep. Invest.*— *U. S. Bur. Mines* **1980,** RI 8480.
3. Matthes, S. A. *Rep. Invest.*—*U. S. Bur. Mines* **1980,** RI 8484.
4. Matthes, S. A.; Farrell, R. F.; Mackie, A. J. *Tech. Prog. Rep.*—*U. S. Bur. Mines* **1983,** TPR 120.
5. Matthes, S. A. Presented at the 35th Pittsburgh Conference, Atlantic City, NJ, 1984, Paper No. 985.
6. Matthes, S. A. *Assoc. Offic. Anal. Chem. Workshop,* 1986 No. 210.
7. Nadkarni, R. A. *Anal. Chem.* **1984,** 56, 2233–2237.
8. Smith, F.; Cousins, B.; Bozic, J.; Flora, W. *Anal. Chim. Acta* **1985,** 177, 243–245.
9. Labrecque, J. Proc. 17th Ann. Can. Min. Anal. Conf., Sept. 1985, pp 1738.
10. Fischer, L. B. *Anal. Chem* **1986,** 58, 261–263.
11. Fernando, L. A.; Heavner, W. D.; Gabrielli, C. C. *Anal. Chem.* **1986,** 58, 511–512.
12. Lamothe, P. J.; Fries, T. L.; Consul, J. J. *Anal. Chem.* **1986,** 58, 1881–1886.
13. Eisenman, M.; Mahaffey, E. Cominco American, private communication, 1986.
14. Bozic, J.; Delvecchio, R. INCO Ltd., private communication, 1986.

RECEIVED for review July 21, 1987. ACCEPTED revised manuscript December 16, 1987.

CHAPTER 4

Open Reflux Vessels for Microwave Digestion

Botanical, Biological, and Food Samples for Elemental Analysis

R. Thomas White, Jr.

"In the fields of observations, chance favors only the mind that is prepared".

Louis Pasteur

The development of two methods of sample preparation using digestion solutions heated by microwave energy is described. The methods tested yield data that compare favorably with certified values on 13 National Bureau of Standards standard reference materials. In the first method, an open (Kohlrausch) vessel containing HNO_3 and H_2O_2 is heated internally by microwave energy to digest organic material. In the second method, a Teflon [poly(tetrafluoroethylene)] vessel and reflux top are used with HNO_3, H_2O_2, and HF to prepare samples for elemental analysis for Al and Si. For each method, samples remain in their original vessels throughout the preparation steps and final volume adjustment. Sample preparation time for elemental analysis can be reduced by 80% by using a microwave digestion system.

AGRICULTURAL LABORATORIES THAT ADVISE FARMERS and growers around the world are testing millions of plant tissue and food product samples each year (1). Generally, this high volume of samples allows only the analysis of the major and minor elements that are essential for plant growth. In large analytical programs, the cost per analysis and sample throughput become valid considerations and justify the use of fast open-vessel digestions for elemental analysis.

Well-documented wet digestion methods that employ open vessels are used for sample preparation for atomic absorption (AA) and inductively coupled plasma (ICP) (2, 3). The open vessel is incorporated in a commercial automated wet digestion device with a heating block. This design is most beneficial for large sample loads. The open-vessel design, however, precludes its use to prepare samples for ultratrace metal analyses (4). Wet digestion methods are among the most widely used methods for the digestion of organic samples for elemental analysis. The elements that are liberated and remain in the sample solution can be predicted fairly accurately (5).

Volatilization of some elements during digestion of organic samples can be a major source of error. For many elements of interest, however, the loss is relatively small and is within acceptable and predictable limits. If the determination of those elements that are subject to volatilization and loss during the decomposition of organic material are of primary importance, then a closed-vessel system should be used.

Previously, preparation of samples for spectrochemical analysis required the use of classical methods. Furnace ashing, heating blocks, and the widely used method of heating sample containers on hot plates involved the slow transfer of energy in the form of heat through sample vessels to warm acid for complete destruction of organic material.

In 1945, a scientist at the Raytheon Company accidentally placed a chocolate bar beside a radar vacuum tube he was testing. Following his discovery, microwave technology was quickly and successfully put to commercial use in 1947 with the construction of one of the earliest commercial microwave units (6). In 1975, researchers at the University of Missouri first described a microwave digestion method as a very safe, rapid, and convenient way to wet ash samples (7). A commercial microwave digestion unit, MDS-81, specifically designed for sample decomposition with acid, was introduced in 1985. The microwave unit, produced by CEM Corporation, encouraged the development of open-vessel digestion methods that use microwave technology for sample preparation that provides an 80% reduction in preparation time when compared to classical methods.

Three major objectives for microwave–acid decomposition of botanical, biological, and food samples for elemental quantitation are safety, speed, and accuracy.

Papers dealing with the development of this new acid decomposition method (6–14) described various vessel designs for microwave heating. Early publications reported the use of open vessel methods that offered speed, accuracy, and the ability to handle large sample loads. Even shorter reaction times and greater efficiency of sample preparation have been developed with closed-vessel systems that provide elemental analysis at major, minor, and trace levels (15–20). Food samples weighing up to 5 g have been prepared by HNO_3–$HClO_4$ mixtures in a microwave system (21). Most recently, a continuous-flow sample and reagent injection system has been described that allows the mineralization of whole blood by microwave energy as it passes through the microwave unit prior to analyses (16).

Experimental Section

Methods have been developed in our laboratory to prepare botanical, biological, and food products with a microwave digestion system. These samples are digested during normal working hours, followed by an unattended, overnight analysis of the sample solution by computer-controlled flow injection inductively coupled plasma emission spectrometry (FIA–ICP).

Flow Injection–Inductively Coupled Plasma Spectrometry

The spectrometer used to analyze microwave-prepared samples was a multichannel 34000 Applied Research Laboratory simultaneous instrument with a DEC PDP 11/03 computer. A reservoir containing 516 gallons of liquid argon provided a continuous supply of argon gas for weeks of unattended instrument operation. The flow injection equipment consisted of a FIAtron microprocessor-controlled SHS-300 with a programmable sample changer. Work was completed on the automated functions of this equipment and reported at both the 1985 Federation of Analytical Chemistry and Spectroscopy Societies Meeting by Martin et al. (*22*) and 1986 Winter Conference on Plasma Spectrochemistry by Dobbins et al. (*23*). They described an intelligent, automatic sample-handling and analysis system with modified software. They sought the automatic analysis of undiluted and stream diluted samples, as well as the automatic dilutions and analysis of samples of widely varying elemental composition and concentration. Samples could be analyzed automatically, and all elements could be quantitated within an optimum range on the calibration curve. The minor elements were read without dilution, and major elements could be read on a variety of samples with dilution, if the computer determined dilution steps were required for accurate analysis.

   Illustration: A sample of standard reference material (SRM) citrus leaves (SRM-1572) is analyzed undiluted. Values for Ca, Mg, K, and S are above the highest calibration standard. The measured concentrations for P, Ba, Mn, Na, and Zn are within the calibration curve range (Table 4-1). The valid values are stored in a temporary file for future recall while the spectrometer is instructed to read the sample again after 10-fold dilution. As seen in the second printout, Mg, K, and S are within range and stored as legitimate values (Table 4-2). Ca is still out of range and causes the computer to direct the instrument to make a 65-fold dilution on the third analysis (Table 4-3). The third printout reveals that Ca is a valid value. The elements of interest are all quantitated within the optimum range on the calibration curve. At this point, the reagent blank is subtracted, sample dilution correction factors are applied, weight–volume calculations are made, and a final report generated (Table 4-4). The final report contains "D" symbols beside

Table 4-1. FIA–ICP Printout of Undiluted Citrus Leaves Sample

Element	Signal mV	Measured Conc.	Actual Conc.
>Ca	12859.53	>106.38	>106.38
>Mg	9829.30	>115.06	>115.06
P	68.97	24.054	24.054
>K	1596.18	>374.68	>374.68
>S	487.47	> 83.207	> 83.207
Ba	53.66	0.3413	0.3413
Mn	63.95	0.4184	0.4184
Na	70.82	3.0208	3.0208
Zn	53.09	0.4991	0.4991

NOTE: Enter wt./vol., 2.0306,100; Flow Injection Sequence 1; Dilution Factor, 1; Time, 15:01:44.
> indicates that the value is above the highest concentration standard.

each element that required dilution. The report lists the final concentration in micrograms per milliliter, and in the right hand column, the final results in micrograms per gram. This report is saved on disk for transfer to a larger computer for permanent storage.

The automated FIA–ICP system developed by Martin et al. (22) and Dobbins et al. (23) is used in this work to analyze the 2.0-g (open vessel) and 1.0-g (reflux vessel) samples prepared by microwave methods.

An analysis by automated FIA–ICP requires 9.5 min per sample, which includes three readings, if required, and a final report. The automatic sample changer has a maximum capacity of 76 samples. A full-capacity run that is

Table 4-2. FIA–ICP Printout of 10-Fold Diluted Citrus Leaves Sample

Element	Signal mV	Measured Conc.	Actual Conc.
>Ca	9625.36	>77.748	>689.93
Mg	1551.14	13.134	D 116.55
P	16.61	2.8505	24.054
K	251.51	40.169	D 356.46
S	71.87	9.7839	D 86.823
<Ba	28.95	0.0375	0.3413
<Mn	18.98	0.0424	0.4184
<Na	43.00	0.2731	3.0208
Zn	20.67	0.0146	0.4991

NOTE: Flow Injection Sequence 2; Dilution Factor, 8.874; Time, 15:04:34; D indicates that the element required dilution.
> indicates that the value is above the highest concentration standard.
< indicates that the value is below the highest concentration standard.

Table 4-3. FIA–ICP Printout of 65-Fold Diluted Citrus Leaves Sample

Element	Signal mV	Measured Conc.	Actual Conc.
Ca	1449.67	10.112	D 658.32
Mg	578.53	1.8968	D 116.55
P	10.49	0.2530	24.054
K	87.59	5.5136	D 356.46
S	24.32	1.6932	D 87.280
<Ba	26.08	< 0.0033	0.3413
<Mn	13.69	< 0.0004	0.4184
<Na	39.75	< 0.0079	3.0208
<Zn	16.98	< 0.0035	0.4991

NOTE: Flow Injection Sequence 3; Dilution Factor, 65.10; Time, 15:07:22.
D indicates that the element required dilution.
< indicates that the value is below the highest concentration standard.

initiated at 5:00 p.m. and allowed to run overnight, is completed by 5:00
a.m. the next day. This sequence can be continued for as long as a week
changing only peristaltic tubing and performing calibration normalizations
between sample loads. The stability of the analytical system and accuracy
of the data are monitored by including a certified reference material every
10th sample.

Microwave Digestion System

A microwave digestion system, Model MDS-81 (CEM Corporation, Indian
Trail, NC) can be used as purchased, to prepare the samples for FIA–ICP
analysis. A clean atmosphere is maintained in the oven cavity during diges-

Table 4-4. FIA–ICP Printout of Final Data for Citrus Leaves

Element	Final Conc. μg/ml	Dilution Corrected Conc. μg/g
D Ca	656.90	32058.1
D Mg	116.27	5674.0
P	24.054	1173.9
D K	356.30	17388.1
D S	87.268	4258.8
Ba	0.3413	16.658
Mn	0.3618	17.655
Na	2.8645	139.79
Zn	0.4750	23.183

NOTE: D indicates that the element required dilution.

tion of samples by adapting a Lab-Aire polyurethane filter (Bel-Art, H18823) at incoming air ports that supply exhaust air. The unit is routinely checked for excessive microwave radiation leakage with a microwave survey meter (Holaday Industries, Model HI 1501).

Acid Scrubber

Safety is of primary importance during the microwave digestion of samples with acid. The fumes generated during the acid decomposition of organic material must be removed from the microwave cavity to a safe location to avoid contaminating the air in the laboratory. Contamination could endanger the analyst and possibly damage the microwave electronics.

Researchers are devising various means for acid fume removal. The removal methods include venting directly into fume hoods (15,20); using scrubbers of water–fiberglass (7, 25), water–gravel (10), and KOH (24, 26); and using vacuum desiccator collectors (9, 13, 14). A high-efficiency fume scrubbing system, Mystaire Model HS-7 (Heat Systems-Ultrasonics, Inc., Plainview, NY) is used in this study to remove acid fumes during sample preparation. The scrubber contains a special 2-in.-thick mesh scrubbing element consisting of layers of woven polypropylene (PVC)-coated fiberglass mesh. Air entering the scrubber is diffused rather than bubbled through water, permitting the intimate mixing of gas and liquid for maximum scrubbing efficiency. The clean gas then passes through an integral mist eliminator that removes any entrained moisture, resulting in a clean, droplet-free exhaust entering a hood (Figure 4.1).

Component Table

An epoxy-painted metal table is constructed to support the acid scrubber above the microwave system. The acid scrubber is mounted on top of the table in view of the operator with the microwave beneath the table for easy access during sample preparation. The entire compact microwave digestion system is placed under a fume hood during the sample preparation period for added safety and convenience. The hood exhaust air keeps any acid fumes from escaping while the analyst removes open vessels from the microwave cavity. The component table (Figure 4.2) concept provides easy access to each section of the compact microwave digestion system to allow routine maintenance service.

Open-Vessel Method—0.5-g Sample

A safe, rapid, and accurate digestion method using HNO_3 and H_2O_2 heated in an open (Kohlrausch) vessel by microwave energy can prepare a 0.5-g sample of plant tissue for determination of the total concentration for nine elements (Ba, Ca, Mg, Mn, P, K, Na, S, and Zn) by ICP spectroscopy (10).

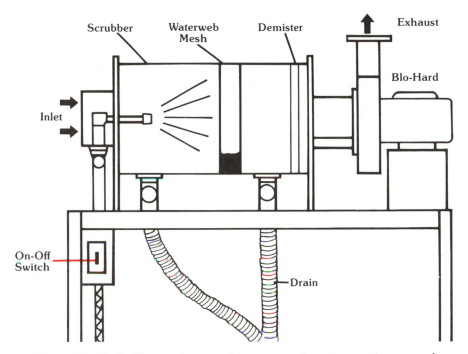

Figure 4.1. *High-efficiency fume scrubbing system for microwave/open vessel acid digestions.*

The elements of interest are present at both major and minor concentrations. The major concentrations require manual dilution to bring their concentrations within the standard calibration range of the instrument. The sample solutions are read on a Perkin-Elmer Model 6000 sequential ICP spectrometer.

A 0.5-g sample prepared in open vessel to a final volume of 100 mL is large enough to determine nine elements for four botanical materials (Table 4-5). Additional samples of agricultural food products and biological materials are also analyzed, and the values are compared to the certified values in the NBS SRMs (Table 4-6).

Open-Vessel Method—2.0-g Sample

Chemists who now use classical methods (e.g., the use of hot plates to heat sample solutions) can take advantage of open-vessel microwave techniques while maintaining standard (up to 2.0-g) sample proportions.

The conditions are determined for the preparation of a 2.0-g sample digested with HNO_3-H_2O_2 in an open (Kohlrausch) flask. (Figure 4.3). A dry 2.0-g sample is weighed directly into a 100-mL Kohlrausch flask. Ten milliliters of HNO_3 is added to the sample vessel so that, without splashing, it washes down any sample material that remains in the flask neck. A Teflon

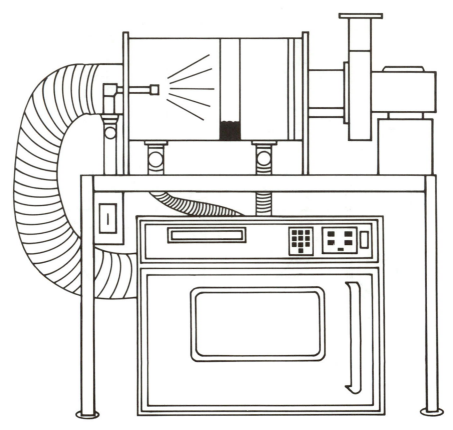

Figure 4.2. Microwave digestion system with fume scrubber and table.

[poly(tetrafluoroethylene) carousel containing 11 samples and a reagent blank is placed into the microwave cavity as a unit (Figure 4.4). The larger 2.0-g sample is started at 50% power for 5 min before increasing to 100% power. The slow start allows the acid to thoroughly wet and predigest the sample material prior to a more vigorous digestion process. During the 100% power step, a balance of refluxing, decomposition, and evaporation is established in the uniquely designed Kohlrausch flask. Near the end, samples are observed through the door and removed when 1–2 mL of HNO_3 remains in the flask. A second 10 mL of HNO_3 and 1 mL H_2O_2 are added to the reaction vessel. The mixture is kept at room temperature for 20 min, or until effervescence has ceased. The carousel of sample flasks is returned to the microwave for the second program step. The samples are heated for 30–40 min and removed when 1–2 mL HNO_3 is still present, to prevent the samples from drying out, charring, and losing volatile elements. The final 5 mL of HNO_3 and 1 mL of H_2O_2 are added to the flask. The extra H_2O_2 oxidizes any remaining organic material, and the effervescence ensures

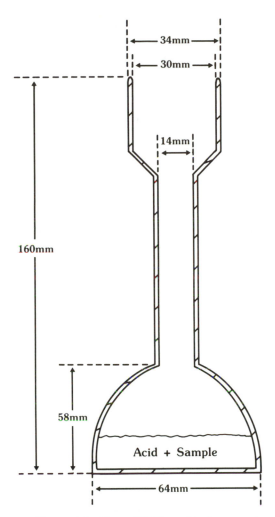

Figure 4.3. Open (Kohlrausch) glass vessel.

that all siliceous material is suspended in the sample solution. The sample solution is diluted to 100 mL with deionized water in the same digestion vessel. After thorough mixing, the sample solution is filtered through a 0.45-μm polycarbonate (Nalgene) filter unit to remove particles of silica that might clog the ICP sample nebulizer. The samples are then loaded into tubes on the programmable sample changer and analyzed overnight by automated FIA–ICP. The step-by-step conditions for the digestion of either a 0.5-g or 2.0-g sample with an open vessel in a microwave digestion system are shown in Table 4-7.

To evaluate the reliability of data from the 2.0-g sample, 13 NBS SRMs were prepared in the microwave unit and analyzed by FIA–ICP with the

Table 4-5. Botanical Samples (0.5 g) Prepared by Microwave Digestion System Compared to Certified Values in NBS Standard Reference Materials

Element	1570 Spinach	1572 Citrus Leaves	1573 Tomato Leaves	1575 Pine Needles
Ca, wt %	1.34 ± 0.07	3.19 ± 0.03	2.91 ± 0.08	0.42 ± 0.01
NBS	1.35 ± 0.03	3.15 ± 0.10	3.00 ± 0.03	0.41 ± 0.02
Mg, wt %	0.86 ± 0.05	0.56 ± 0.01	0.65 ± 0.03	0.12 ± 0.01
NBS	—[a]	0.58 ± 0.03	(0.7)[b]	—
P, wt %	0.53 ± 0.03	0.13 ± 0.00	0.33 ± 0.02	0.12 ± 0.01
NBS	0.55 ± 0.02	0.13 ± 0.02	0.34 ± 0.02	0.12 ± 0.02
K, wt %	3.60 ± 0.06	1.84 ± 0.03	4.34 ± 0.18	0.39 ± 0.02
NBS	3.56 ± 0.03	1.82 ± 0.06	4.46 ± 0.03	0.37 ± 0.02
S, wt %	0.48 ± 0.02	0.459 ± 0.007	0.69 ± 0.03	0.14 ± 0.01
NBS	—	0.407 ± 0.009	—	—
Ba, µg/g	13.9 ± 0.7	24 ± 1	66 ± 3	7.8 ± 4
NBS	—	21 ± 3	—	—
Mn, µg/g	165 ± 10	21 ± 1	219 ± 7	693 ± 6
NBS	165 ± 6	23 ± 2	238 ± 7	675 ± 15
Na, µg/g	15470 ± 212	164 ± 13	369 ± 16	26 ± 9
NBS	—	160 ± 20	—	—
Zn, µg/g	50 ± 3	28 ± 1	61 ± 3	72 ± 13
NBS	50 ± 2	29 ± 2	62 ± 6	—

SOURCE: Reprinted with permission from ref. 10. Copyright 1985 Association of Official Analytical Chemists.
NOTE: All values are the average of 10 replicates.
[a]No listed NBS value.
[b]Noncertified value.

HNO_3-H_2O_2 decomposition method. The botanical SRMs included spinach (SRM-1570), orchard leaves (SRM-1571), citrus leaves (SRM-1572), tomato leaves (SRM-1573), pine needles (SRM-1575), and corn stalk (SRM-8112) (Table 4-8). Biological materials prepared by open-vessel digestion heated with microwave energy included oyster tissue (SRM-1566), brewers yeast (SRM-1569), and bovine liver (SRM-1577a) (Table 4-9). NBS SRMs of agricultural food products prepared by the microwave method included nonfat milk powder (SRM-1549), wheat flour (SRM-1567), rice flour (SRM-1568) and corn kernel (SRM-8413) (Table 4-10). The values for the SRMs prepared by the two open-vessel methods are in good agreement with certified data listed for the SRMs. The only exceptions are the low values for Na for the 2.0-g samples of orchard leaves (SRM-1571) and corn stalk (SRM-8112), and the high value for rice Flour (SRM-1568). The 0.5-g samples are diluted manually and analyzed by sequential ICP. The 2.0-g samples are analyzed overnight by automatic computer-intelligent dilutions with FIA–simultaneous ICP. The analytical values from the microwave-prepared 0.5-g samples are as good as values obtained from unattended analysis of the 2.0-g samples.

Table 4-6. Agricultural Food Products and Biological Samples (0.5 g) Prepared by Microwave Digestion System Compared to Values in NBS Standard Reference Materials

Element	1566 Oyster Tissue	1567 Wheat Flour	1568 Rice Flour	1577 Bovine Liver
Ca, wt %	0.15 ± 0.01	0.017 ± 0.002	0.012 ± 0.003	87 ± 13
NBS	0.15 ± 0.02	0.019 ± 0.001	0.014 ± 0.002	(123)[a]
Mg, wt %	0.131 ± 0.002	0.037 ± 0.002	0.045 ± 0.002	616 ± 19
NBS	0.128 ± 0.009	—[b]	—	(605)
P, wt %	0.77 ± 0.01	0.015 ± 0.000	0.168 ± 0.004	1.24 ± 0.02
NBS	(0.81)	—	—	—
K, wt %	0.963 ± 0.031	0.113 ± 0.019	0.097 ± 0.016	1.05 ± 0.02
NBS	0.969 ± 0.005	0.136 ± 0.004	0.112 ± 0.002	0.97 ± 0.06
S, wt %	0.96 ± 0.02	0.186 ± 0.005	0.140 ± 0.000	0.93 ± 0.01
NBS	(0.76)	—	—	—
Ba, μg/g	NDA[c]	NDA	NDA	NDA
NBS	—	—	—	—
Mn, μg/g	17.9 ± 1.3	7.2 ± 1.0	20.7 ± 1.4	9.9 ± 0.9
NBS	17.5 ± 1.2	8.5 ± 0.5	20.1 ± 0.4	10.3 ± 1.0
Na, μg/g	[0.53 ± 0.1][d]	<20	<20	[0.272 ± 0.019]
NBS	[0.51 ± 0.03]	8.0 ± 1.5	6.0 ± 1.5	[0.243 ± 0.013]
Zn, μg/g	824 ± 9	10.0 ± 0.1	20.4 ± 0.9	153 ± 2
NBS	852 ± 14	10.6 ± 1.0	19.4 ± 1.0	130 ± 10

SOURCE: Reprinted with permission from ref. 10. Copyright 1985 Association of Official Analytical Chemists.
NOTE: All values are the average of 10 replicates.
[a]Noncertified value.
[b]No listed NBS value.
[c]No detectable amount.
[d]wt %.

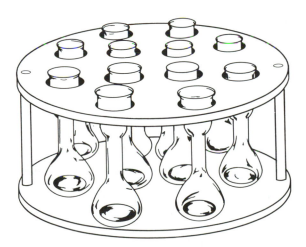

Figure 4.4. Teflon carousel loaded with 12 Kohlrausch vessels.

Table 4-7. Conditions for Microwave/Open Vessel (Kohlrausch Flask) Sample Decomposition

Step	Conditions	0.5-g Sample	2.0-g Sample
1	% Microwave power	90	50
	Heating time (min)	30	5
	% Microwave power	—	100
	Heating time (min)	—	15
	Digestion reagents	10 mL HNO_3	10 mL HNO_3
2	% Microwave power	90	100
	Heating time (min)	30	40
	Digestion reagents	10 mL HNO_3	10 mL HNO_3
		1 mL H_2O_2	1 mL H_2O_2
3	Final reagent	5 mL HNO_3	5 mL HNO_3
		—	1 mL H_2O_2
	Final volume	100 mL	100 mL
4	Filter	0.45 μm	0.45 μm
5	Analyze	ICP	ICP

The 80% reduction in sample preparation time (in comparison to hot plate methods) by microwave digestion combined with automated FIA–ICP will help reduce reduce large sample backlogs. A laboratory equipped with two microwave digestion systems can prepare 24 samples per hour to easily load the ICP autosampler for overnight analysis.

Teflon Reflux Vessel–1.0-g Sample

The vessels selected for microwave decomposition of various samples take many forms. Open-vessel designs have evolved directly to closed-vessel designs. This investigation, however describes a semiclosed reflux vessel that bridges the gap between open and closed vessels. The vessel was designed specifically for those analysts seeking additional flexibility and speed in rapid Al and Si analysis. The requirements for a reflux vessel are

- Teflon construction
- Cone top to stimulate condensing/refluxing action
- Rapid disassembly of container sections
- Dimensions to maximize sample containers per digestion
- 1-g sample capacity
- HF resistant vessel material

The vessel and carousel are made of Teflon PTFE purchased from a commercial supplier of Teflon labware (Berghof/America, Inc. Raymond,

Table 4-8. Botanical Samples (2.0 g) Prepared by Open (Kohlrausch) Digestion Vessel Microwave Method Compared to Certified Values in NBS Standard Reference Materials

Element	1570 Spinach[a]	1571 Orchard Leaves	1572 Citrus Leaves	1573 Tomato Leaves	1575 Pine Needles	8412 Corn Stalk
Ca, wt %	1.27	2.05 ± 0.03	3.25 ± 0.10	2.99 ± 0.14	0.42 ± 0.01	0.217 ± 0.008
NBS	1.35 ± 0.03	2.09 ± 0.03	3.15 ± 0.10	3.00 ± 0.03	0.41 ± 0.02	0.216 ± 0.008
Mg, wt %	0.79	0.58 ± 0.02	0.58 ± 0.02	0.64 ± 0.03	0.11 ± 0.00	0.154 ± 0.006
NBS	—[b]	0.62 ± 0.02	0.58 ± 0.03	(0.7)[c]	—	0.160 ± 0.007
P, wt %	0.57	0.20 ± 0.00	0.13 ± 0.00	0.34 ± 0.02	0.13 ± 0.00	0.06 ± 0.00
NBS	0.55 ± 0.02	0.21 ± 0.01	0.13 ± 0.02	0.34 ± 0.02	0.12 ± 0.02	0.00
K, wt %	3.02	1.42 ± 0.04	1.74 ± 0.06	4.25 ± 0.19	0.35 ± 0.01	1.62 ± 0.05
NBS	3.56 ± 0.03	1.47 ± 0.03	1.82 ± 0.06	4.46 ± 0.03	0.37 ± 0.02	1.73 ± 0.05
S, wt %	0.41	0.17 ± 0.00	0.451 ± 0.014	0.67 ± 0.03	0.13 ± 0.01	0.066 ± 0.002
NBS	—	(0.19)	0.407 ± 0.009	—	—	—
Ba, µg/g[d]	12	41 ± 1	20 ± 0	52 ± 1	7 ± 0	5 ± 0
NBS	—	(44)	21 ± 3	—	—	—
Mn, µg/g	152	84 ± 1	20 ± 1	215 ± 3	642 ± 8	14 ± 0[e]
NBS	165 ± 6	91 ± 4	23 ± 2	238 ± 7	675 ± 15	15 ± 2
Na, µg/g	12397	34 ± 1	140 ± 4	302 ± 15	13 ± 3	18 ± 4
NBS	—	82 ± 6	160 ± 20	—	—	28 ± 8
Zn, µg/g[d]	36 ± 3	22 ± 1	26 ± 1	57 ± 2	74 ± 6	30 ± 5
NBS	50 ± 2	25 ± 3	29 ± 2	62 ± 6	—	32 ± 3

NOTE: All values are the average of 11 replicates.
[a] Single sample.
[b] No listed NBS value.
[c] Noncertified value.
[d] Sequential ICP.

Table 4-9. Biological Samples (2.0 g) Prepared by Open (Kohlrausch) Digestion Vessel Microwave Method Compared to Certified Values in NBS Standard Reference Materials

Element	1566 Oyster Tissue	1569 Brewers Yeast	1577 Bovine Liver
Ca, wt %	0.14 ± 0.00	0.230 ± 0.003	[118 ± 3][a]
NBS	0.15 ± 0.02	—[b]	[120 ± 7]
Mg, wt %	0.118 ± 0.002	0.149 ± 0.002	[612 ± 6]
NBS	0.128 ± 0.009	—	[600 ± 15]
P, wt %	0.87 ± 0.03	1.07 ± 0.03	1.32 ± 0.03
NBS	(0.81)[c]	—	1.11 ± 0.04
K, wt %	0.931 ± 0.019	1.47 ± 0.04	1.042 ± 0.020
NBS	0.969 ± 0.005	—	0.996 ± 0.007
S, wt %	0.90 ± 0.02	0.44 ± 0.01	0.90 ± 0.01
NBS	(0.76)	—	0.78 ± 0.01
Ba, μg/g[d]	4.2 ± 0.3	1.5 ± 0.3	2.1 ± 0.3
NBS	—	—	—
Mn, μg/g[d]	16.3 ± 0.4	7.9 ± 0.3	10.2 ± 0.3
NBS	17.5 ± 1.2	—	9.9 ± 0.8
Na, μg/g	[0.48 ± 0.01][e]	92.5 ± 2.8	[0.227 ± 0.002]
NBS	[0.51 ± 0.03]	—	[0.243 ± 0.013]
Zn, μg/g	774 ± 10	58 ± 2	109 ± 2
NBS	852 ± 14	—	123 ± 8

NOTE: All values are the average of 11 replicates.
[a] μg/g.
[b] No listed NBS value.
[c] Noncertified value.
[d] Sequential ICP.
[e] wt %.

NH). The vessel (#140-03-50) with thick-wall and rigid construction is used as purchased. A round-bottom Teflon tube (#14012-06) serves as a reflux cone for the vessel. A small amount of lathe turning produces an insert to fit inside the vessel with an edge for cone support (Figure 4.5). A carousel is constructed with slots to hold 21 vessels for placement in the microwave cavity (Figure 4.6).

The determination of total Al and Si in both biological and botanical materials and also food products requires either a flux–fusion method or use of HF. Fusion in metal crucibles with fluxes may lead to incomplete attack of the sample, limited solubility of some metal ions in the particular flux environment, more extensive sample contamination, and losses caused by reduction and alloying. The high salt concentrations from a 10- to 20-fold excess of flux weight over sample weight, cause solution instability, fluctuating instrument background readings, and clogged nebulizers (27).

Table 4-10. Agricultural Food Product (2.0 g) Samples Prepared by Open (Kohlrausch) Digestion Vessel Microwave Method Compared to Certified Values in NBS Standard Reference Materials

Element	1549 Nonfat Milk Powder	1567 Wheat Flour	1568 Rice Flour	8413 Corn Kernel
Ca, wt %	1.36 ± 0.05	0.019 ± 0.001	0.015 ± 0.001	[42 ± 4][a]
NBS	1.30 ± 0.05	0.019 ± 0.001	0.014 ± 0.002	[42 ± 5]
Mg, wt %	0.117 ± 0.004	0.036 ± 0.001	0.042 ± 0.001	[1000 ± 10]
NBS	0.120 ± 0.003	—[b]	—	[990 ± 82]
P, wt %	1.08 ± 0.05	0.16 ± 0.00	0.18 ± 0.01	0.23 ± 0.01
NBS	1.06 ± 0.02	—	—	—
K, wt %	1.70 ± 0.07	0.122 ± 0.001	0.103 ± 0.002	[3640 ± 60]
NBS	1.69 ± 0.03	0.136 ± 0.004	0.112 ± 0.002	[3570 ± 370]
S, wt %	0.390 ± 0.011	0.166 ± 0.002	0.124 ± 0.004	0.115 ± 0.002
NBS	0.351 ± 0.005	—	—	—
Ba, μg/g[e]	3.2 ± 0.4	1.5 ± 0.2	NDA[c]	1.2 ± 0.2
NBS	—	—	—	—
Mn, μg/g[e]	0.33 ± 0.10	7.7 ± 0.2	19.1 ± 0.2	4.2 ± 0.3
NBS	0.26 ± 0.06	8.5 ± 0.5	20.1 ± 0.4	4.0 ± 0.3
Na, μg/g	[0.479 ± 0.020][d]	8.0 ± 3.4	9.1 ± 1.7	NDA
NBS	[0.497 ± 0.010]	8.0 ± 1.5	6.0 ± 1.5	—
Zn, μg/g[e]	44.2 ± 0.5	9.1 ± 0.3	17.3 ± 0.8	14.3 ± 1
NBS	46.1 ± 2.2	10.6 ± 1.0	19.4 ± 1.0	15.7 ± 1.4

NOTE: All values are the average of three replicates.
[a]μg/g.
[b]No listed NBS value.
[c]No detectable amount.
[d]wt %.
[e]Sequential ICP.

Using a Teflon reflux vessel for the analysis of Al and Si allows the use of HNO_3–H_2O_2–HF in the decomposition steps, eliminates the need for salt fluxes, and overcomes many of the difficulties and disadvantages encountered with the flux system.

A sample (1.0-g) may be weighed directly into the sample container. Ten milliliters HNO_3 is added to the reaction vessel, washing down the sides and completely wetting the sample. A total of 20 samples and a reagent blank are placed in the Teflon carousel, which is loaded as a unit into the microwave cavity. Predigestion of the sample begins with the microwave power set at 50% for 5 min before increasing to 100% power for 20–30 min. The microwave power is maintained at 100% for 20–40 min during the second ashing step. The first and second steps of sample decomposition are conducted with the reflux cone off; this allows complete digestion of the organic material by HNO_3–H_2O_2. In the third step, the final 5 mL of HNO_3 and 1 mL of H_2O_2 are added to the Teflon vessel. The mixture is kept at room temperature until effervescence has ceased. HF (1.5 mL/50 mL) is

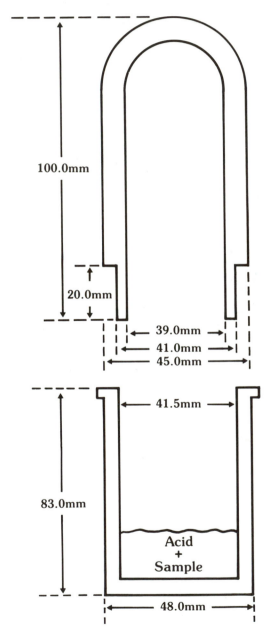

Figure 4.5. Teflon decompostion vessel and reflux top.

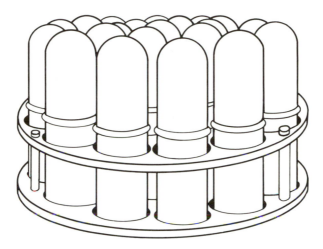

Figure 4.6. Teflon carousel loaded with 21 reflux vessels.

then added to dissolve the siliceous material. A Teflon reflux top placed on each vessel provides a sufficient seal for gentle heating and refluxing of acid at 30% power for 30 min with the sample. Finally, 43.5 mL of H_3BO_3 solution (1.4 g/50 mL) is added with an automatic dispenser. The final volume in the vessel is 50 mL. The sample vessel, with reflux top, is returned to the microwave. The sample solution is then warmed at 30% power to complex the fluoride and to dissolve any precipitated metal fluorides. The sample solution is filtered through a 0.45-μm filter and is analyzed. There is no danger of HF damage to the ICP quartz spray chamber or nebulizer (Table 4-11).

The reflux vessel method is evaluated with the same 13 NBS SRMs that were used to evaluate the open (Kohlrausch) vessel. Nine elements are determined on the six botanical SRMs: spinach, orchard leaves, citrus leaves, tomato leaves, pine needles and corn stalk (Table 4-12). The three biological samples of oyster tissue, brewers yeast and bovine liver were also analyzed by automated FIA–ICP (Table 4-13). Finally, nonfat milk powder, wheat flour, rice flour, and corn kernel containing certified elements are the four agricultural food products tested by the method (Table 4-14).

The values obtained by the reflux vessel method are within the certified range or lower than the values for the 13 SRMs. The erratic behavior of Na and the low values for Mg and Zn read on undiluted sample solutions are unexplained. A low value was obtained for Mn in the nonfat milk powder (SRM-1549) , but adequate Mn values were obtained for the 11 additional certified values listed on the SRMs analyzed. Nutrient data for Al and Si, as well as for additional elements, allowed farmers and growers to be advised of crop conditions quickly. A microwave sample preparation cycle for a carousel of 21 reflux vessels can be prepared in less than 2 h as compared

Table 4-11. Conditions for Microwave/Teflon Reflux
Vessel Sample Decomposition

Step	Conditions	1.0-g Sample
1	Reflux top	"Off"
	% Microwave power	50
	Heating time (min)	5
	% Microwave power	100
	Heating time (min)	20-30
	Digestion reagents	10 mL HNO_3
2	Reflux top	"Off"
	% Microwave power	100
	Heating time (min)	20-40
	Digestion reagents	10 mL HNO_3
		1 mL H_2O_2
3	Reflux top	"On"
	% Microwave power	30
	Heating time (min)	30
	Final reagents	5 mL HNO_3
		1 mL H_2O_2
		1.5 mL HF
4	Reflux top	"On"
	% Microwave power	30
	Final reagents	43.5 mL H_3BO_3
5	Filter	0.45 μm
6	Analyze	ICP

to 10 h for the flux fusion of an equal number of samples in expensive platinum crucibles. A comparison of the Teflon reflux vessel method to the open glass vessel methods demonstrates the advantage of preparation and quatitation of total Al and Si using the same sample. Because the open vessel is made of glass, it may contaminate the mixture and prevent the accurate determination of Al and Si.

The Al and Si data are obtained by both the HNO_3–H_2O_2–HF method and a borate-flux fusion of the 13 NBS reference materials. There was reasonable agreement with certified values except at low concentrations. In addition to the disadvantages for the flux-fusion method, platinum crucibles and automatic fluxers when used, can exceed the cost of Teflon vessels and microwave equipment for analyses of Al and Si (Table 4-15).

Both classical wet digestion methods and decomposition reagents can be compared to the open- and reflux-vessel microwave methods. The digestion of 0.5–2.0-g samples in 12 open glass vessels with HNO_3–H_2O_2 heated by microwave energy requires 1 h. In a 12-tube aluminum block digestor, 0.5–1.0-g samples require nearly 5 h when using HNO_3–$HClO_4$ to digest the organic material. Increased safety precautions and specially designed hood equipment are required when working with $HClO_4$. Samples digested

Table 4-12. Botanical Samples (1.0 g) Prepared by Teflon Reflux Digestion Vessel Microwave Method Compared to Certified Values in NBS Standard Reference Material

Element	1570 Spinach	1571 Orchard Leaves	1572 Citrus Leaves	1573 Tomato Leaves	1575 Pine Needles	8412 Corn Stalk
Ca, wt %	1.24 ± 0.03	1.87 ± 0.02	2.94 ± 0.03	2.79 ± 0.06	0.39 ± 0.01	0.195 ± 0.003
NBS	1.35 ± 0.03	2.09 ± 0.03	3.15 ± 0.10	3.00 ± 0.03	0.41 ± 0.02	0.216 ± 0.008
Mg, wt %	0.77 ± 0.06	0.46 ± 0.04	0.36 ± 0.01	0.60 ± 0.00	0.10 ± 0.00	0.117 ± 0.009
NBS	—[a]	0.62 ± 0.02	0.58 ± 0.03	(0.7)[b]	—	0.160 ± 0.007
P, wt %	0.46 ± 0.00	0.18 ± 0.00	0.12 ± 0.00	0.29 ± 0.01	0.10 ± 0.00	0.047 ± 0.001
NBS	0.55 ± 0.02	0.21 ± 0.01	0.13 ± 0.02	0.34 ± 0.02	0.12 ± 0.02	—
K, wt %	3.46 ± 0.02	1.31 ± 0.01	1.61 ± 0.02	4.19 ± 0.08	0.34 ± 0.01	1.56 ± 0.03
NBS	3.56 ± 0.03	1.47 ± 0.03	1.82 ± 0.06	4.46 ± 0.03	—	—
S, wt %	0.44 ± 0.00	0.16 ± 0.00	0.401 ± 0.010	0.59 ± 0.01	0.11 ± 0.00	0.059 ± 0.001
NBS	—	(0.19)	0.407 ± 0.009	—	0.11 ± 0.00	0.059 ± 0.001
Ba, µg/g[c]	15 ± 0	38 ± 2	13 ± 1	58 ± 3	6.9 ± 0.2	4.9 ± 0.8
NBS	—	(44)	21 ± 3	—	—	—
Mn, µg/g[c]	154 ± 3	79 ± 3	19 ± 0	207 ± 7	587 ± 6	13 ± 0
NBS	165 ± 6	91 ± 4	23 ± 2	238 ± 7	675 ± 15	15 ± 2
Na, µg/g	13261	63 ± 1	130 ± 5	550 ± 4	61 ± 2	52 ± 2
NBS	—	82 ± 6	160 ± 20	—	—	28 ± 8
Zn, µg/g[c]	37 ± 2	14 ± 1	16 ± 1	55 ± 2	51 ± 5	26 ± 1
NBS	50 ± 2	25 ± 3	29 ± 2	62 ± 6	—	32 ± 3

NOTE: All values are the average of three replicates, except for spinach, which are the average of two replicates.
[a] No listed NBS value.
[b] Noncertified value.
[c] Sequential ICP.

Table 4-13. Biological Samples (1.0 g) Prepared by Teflon Reflux Digestion Vessel Microwave Method Compared to Certified Values in NBS Standard Reference Materials

Element	1566 Oyster Tissue	1569 Brewers Yeast	1577 Bovine Liver
Ca, wt %	0.14	0.222 ± 0.006	$[99 \pm 1]^a$
NBS	0.15 ± 0.02	—[b]	$[120 \pm 7]$
Mg, wt %	0.104	0.158 ± 0.004	$[498 \pm 8]$
NBS	0.128 ± 0.009	—	$[600 \pm 15]$
P, wt %	0.68	0.90 ± 0.01	1.01 ± 0.01
NBS	$(0.81)^c$	—	1.11 ± 0.04
K, wt %	0.863	1.34 ± 0.02	0.894 ± 0.012
NBS	0.969 ± 0.005	—	
S, wt %	0.80	0.40 ± 0.00	0.74 ± 0.01
NBS	(0.76)	—	0.78 ± 0.01
Ba, $\mu g/g^d$	4.4	7.6 ± 0.1	0.7 ± 0.1
NBS	—	—	—
Mn, $\mu g/g^d$	16.0	9.4 ± 0.1	8.6 ± 0.1
NBS	17.5 ± 1.2	—	9.9 ± 0.8
Na, $\mu g/g$	$[0.45]^e$	NDA[f]	$[0.203 \pm 0.013]$
NBS	$[0.51 \pm 0.03]$	—	$[0.243 \pm 0.013]$
Zn, $\mu g/g$	740	62.9 ± 2.9^d	106 ± 1
NBS	852 ± 14	—	123 ± 8

NOTE: All values are the average of three replicates, except for oyster tissue, which are for a single sample.
[a]$\mu g/g$.
[b]No listed NBS value.
[c]Noncertified value.
[d]Sequential ICP.
[e]wt %.
[f]No detectable amount.

with HNO_3–$HClO_4$ in an aluminum heating block cannot be seen during heating. To visually observe the digestion progress, the entire sample tray containing the tubes of boiling acid must be lifted up. The microwave digestion system eliminates this problem because it contains a transparent door to allow observation of the boiling sample solutions.

Samples can be prepared in the microwave digestion system by using aqua regia (1:3 HNO_3:HCl) to decompose the organic material. It takes 1 h to prepare 12 samples by either the microwave digestion system with aqua regia or by the HNO_3-H_2O_2 method. Either method provides equally good results when compared to the other procedures; however, additional steps are required to prepare enough aqua regia for each day (Table 4-16).

Table 4-14. Agricultural Food Product (1.0 g) Samples Prepared by Teflon Reflux Digestion Vessel Microwave Method Compared to Certified Values in NBS Standard Reference Materials

Element	1549 Nonfat Milk Powder	1567 Wheat Flour	1568 Rice Flour	8413 Corn Kernel
Ca, wt %	1.11 ± 0.04	0.016 ± 0.006	0.011 ± 0.000	[45 ± 5][a]
NBS	1.30 0.05	0.019 ± 0.001	0.014 ± 0.002	[42 ± 5]
Mg, wt %	0.092 ± 0.004	0.033 ± 0.000	0.029 ± 0.009	[390 ± 120]
NBS	0.120 ± 0.003	—[b]	—	
			[990 ± 82]	
P, wt %	0.85 ± 0.03	0.12 ± 0.00	0.14 ± 0.00	0.19 ± 0.00
NBS	1.06 ± 0.02	—	—	—
K, wt %	1.42 ± 0.0	0.112 ± 0.001	0.094 ± 0.001	[3300 ± 150]
NBS	1.69 ± 0.03	0.136 ± 0.004	0.112 ± 0.002	[3570 ± 370]
S, wt %	0.331 ± 0.010	0.144 ± 0.001	0.108 ± 0.002	0.101 ± 0.002
NBS	0.351 ± 0.005	—	—	—
Ba, μg/g[c]	1.3 ± 0.1	1.1 ± 0.2	0.3 ± 0.1	2.4 ± 0.0
NBS	—	—	—	—
Mn, μg/g[c]	0.14+ 0.02	7.3 ± 0.1	18.0 ± 0.3	4.6 ± 0.1
NBS	0.26 ± 0.06	8.5 ± 0.5	20.1 ± 0.4	4.0 ± 0.3
Na, μg/g	[0.408 ± 20][d]	29.5 ± 0.5	28.2 ± 1.1	NDA[e]
NBS	[0.497 ± 0.010]	8.0 ± 1.5	6.0 ± 1.5	—
Zn, μg/g[c]	34.0+ 1.2	NDA	7.8 ± 0.6	11.8 ± 0.4
NBS	46.1 ± 2.2	10.6 ± 1.0	19.4 ± 1.0	15.7 ± 1.4

NOTE: All values are the average of three replicates.
[a] μg/g.
[b] No listed NBS value.
[c] Sequential ICP.
[d] wt %.
[e] No detectable amount.

Summary

A safe, dependable, and fast alternative is available to the analyst that presently depends on slow, classical sample preparation methods for elemental analysis. Depending on the specific analysis desired, the analyst can select from either the open (Kohlrausch) vessel or semiclosed Teflon reflux vessel. The open-vessel method can prepare 0.5–2.0-g samples with HNO_3–H_2O_2 for quantitative analysis of nine elements. The Teflon reflux vessel allows rapid preparation of 1.0-g samples with HNO_3–H_2O_2–HF for the analysis for Al and Si.

Table 4-15. Aluminum and Silicon Data by Teflon Reflux Digestion Vessel and Flux Fushion Methods Compared to Certified Values in NBS Standard Reference Material

Reference Material	NBS Certified Values		Microwave HNO$_3$-H$_2$O$_2$-HF		Borate Flux Fusion LiCO$_3$–Boric Acid	
	Al, µg/g	Si, wt %	Al, µg/g	Si/ wt %	Al, µg/g	Si/ wt %
1570 Spinach	870 ± 50	—[a]	852 ± 13	0.34 ± 0.01	—[b]	—
1571 Orchard Leaves	—	—	357 ± 6	0.17 ± 0.01	436	0.22
1572 Citrus Leaves	92 ± 15	—	81 ± 2	0.14 ± 0.01	92	0.15
1573 Tomato Leaves	(1200)[c]	—	1260 ± 1	1.12 ± 0.02	1407	1.16
1575 Pine Needles	545 ± 30	—	551 ± 9	0.13 ± 0.00	724	0.23
8412 Corn Stalk	—	—	86 ± 6	0.27 ± 0.01	97	0.26
1566 Oyster Tissue	—	—	241	0.14	251	0.16
1569 Brewers Yeast	—	—	2066 ± 66	1.28 ± 0.03	2148	1.30
1577a Bovine Liver	(2)	—	NDA[d]	0.04 ± 0.01	3	NDA
1549 Nonfat Milk Powder	(2)	[50][e]	3 ± 0	0.06 ± 0.01	8	0.0001
1567 Wheat Flour	—	—	3 ± 0	NDA	18	0.002
1568 Rice Flour	—	—	NDA	0.001	4	0.009
8413 Corn Kernel	4 ± 2	—	NDA	NDA	13	0.004

NOTE: All values for reflux digestion are the average of 11 replicates; values for flux fusion are for a single sample.
[a]No listed NBS value.
[b]No available sample.
[c]Non-certified value.
[d]No detectable amount.
[e]µg/g.

Table 4-16. Analytical Data from NBS Standard Reference Material 1572 Citrus Leaves Prepared by Four Different Acid Digestion Methods

Element	NBS Certified Value	Heating Block HNO$_3$ + HClO$_4$	Microwave Aqua Regia	Microwave HNO$_2$ + H$_2$O$_2$	Microwave HNO$_3$ + H$_2$O$_2$ + HF
Ca, wt %	3.15 ± 0.10	3.25 ± 0.05	3.28 ± 0.07	3.25 ± 0.10	2.94 ± 0.03
Mg, wt %	0.58 ± 0.03	0.56 ± 0.01	0.56 ± 0.01	0.58 ± 0.02	0.36 ± 0.01
P, wt %	0.13 ± 0.02	0.15 ± 0.00	0.13 ± 0.00	0.13 ± 0.00	0.12 ± 0.00
K, wt %	1.82 ± 0.06	1.67 ± 0.07	1.78 ± 0.03	1.74 ± 0.06	1.61 ± 0.02
S, wt %	0.407 ± 0.009	0.490 ± 0.030	0.415 ± 0.008	0.451 ± 0.014	0.401 ± 0.010
Ba, µg/g[a]	21 ± 3	20 ± 1	19 ± 0	20 ± 0	12 ± 1
Mg, µg/g	23 ± 2	21 ± 0	20 ± 0	20 ± 1	18 ± 1
Na, µg/g	160 ± 20	175 ± 6	148 ± 2	140 ± 4	130 ± 5
Zn, µg/g[a]	29 ± 2	25 ± 1	15 ± 1	26 ± 1	13 ± 1

NOTE: All values are the average of 11 replicates, except for the microwave with HF digestion, which are for three replicates.

[a]Sequential ICP.

The many advantages that the analytical chemist will experience with the microwave methods include:

- an 80% reduction in digestion time
- instantaneous interaction between electromagnetic waves and digestion acid
- easy observation of the sample during digestion
- increased selection of materials for reaction vessels
- prevention of contamination of samples by filtering air for exhaust and by using a noncontaminating Teflon cavity
- reduction in the need for HNO_3–$HClO_4$–H_2SO_4 digestion (and therefore reduction in the formation of insoluble salts)
- quick and safe removal of acid fumes with an acid scrubber
- a complimentary digestion method to an automatic plasma spectrometer

The 80% reduction in sample preparation time and the ability to instantly begin interaction of microwave energy with digestion reagents can drastically change the present sample preparation methods for elemental analysis.

The coupling of rapid microwave decomposition methods with an automated simultaneous spectrometer is a state-of-the-art solution for analyzing large numbers of samples. This system provides an opportunity for the analyst to perform the analytical determinations as well as dilutions using unattended computer-controlled flow-injection ICP.

With the present surge of robotics into the modern-day laboratory, a robot-controlled microwave digestion system capable of unattended sample preparation is a welcomed addition to this present automated analysis system.

Acknowledgment

I thank John M. Martin for both technical assistance and helpful discussions during the preparation of this manuscript. The computerized literature survey was completed with the assistance of Helen S. Chung and Christa N. Mathis. I gratefully acknowledge the contributions of Jan S. Riley in the preparation of this manuscript. I especially thank my wife, Sara I. White, for her contributions and support throughout the duration of this project.

Literature Cited

1. Jones, B. G., Jr. In *Developments in Atomic Plasma Spectrochemical Analysis*; Barnes, R. M. Ed.; Heyden: 1981; pp 644–645.
2. Krishnamurtz, K. V.; Skpirt, E.; Redding, M. M. *At. Abs. Newslett.* **1976**, *15*, 68–70.
3. Dahlquist, R. L.; Knoll, J. W. *Appl. Spectrosc.* **1978**, *32*, 1–29.
4. Tatro, M. E. *Spectrosc.* **1986**, *1*, 20.
5. Knapp, G.; *ICP Info. Newslett.*; Barnes, R. M. Ed.; **1984**, *10*, 4.
6. Scott, L. M.; Scott, J.D. *Mastering Microwave Cooking*; Bantam: New York, 1979; p 2.
7. Abu-Samra, A.; Morris, J. S.; Koirtyohann, S. R. *Anal. Chem.* **1975**, *47*, 1476.
8. Schiffmann, R. F. *Food Product Development* **1979**, *13*, 38.
9. Nadkarni, R. A. *Anal. Chem.* **1984**, *56*, 2234.
10. White, R. T. Jr.; Douthit, G. E. *J. Assoc. Off. Anal. Chem.* **1985**, *68*(4), 766–769.
11. Tsukada, S.; Demura, R.; Yamamoto, I. *Eisei Kagaku* **1985**, *31*, 38.
12. Demura, R.; Tsukada S.; Yamamato, I. *Eisei Kagaku* **1985**, *31*, 406.
13. Willging, E. M.; Maroney, G. M.; Garza, T.; Mahan, K. I. Presented at the ACS/SAS Pacific Conference on Chemistry and Spectroscopy, San Francisco, CA, Oct. 1985.
14. Blust, R.; Van de Linden, A.; Decleir, W. *At. Spectrosc.* **1985**, 6, 163–164.
15. Jassie, L. B.; Kingston, H. M. 1985 Pittsburgh Conference Abstracts, Paper No. 108A.
16. Burguera, M.; Burguera, J. L. *Analytica Chemica Acta* **1986,** *179*, 351–352.
17. Fernando, L. A.; Heavner, W. D.; Gabrielli, C. C. *Anal. Chem.* **1986**, 58, 511–512.
18. Fischer, L. B. *Anal. Chem.* **1986**, 58, 261–263.
19. Westbrook, W. T.; Jefferson, R. H. *J. Microwave Power* **1986,** *21*, 29.
20. Matthes, S. A.; Farrell, R. F.; Mackie, A. J. Tech. *Progr. Rep. U. S. Bur. Mines* **1983**, No. 120.
21. Ishu, M. *Mukogawa Joshi Diagaku Kiyo* **1984**, *32*, 1–8.
22. Martin, J. M.; Dobbins, J. T. Jr.; Ihrig, P.J. 1985 Federation of Analytical Chemistry and Spectroscopy Societies Meeting Abstracts, Paper 3242.
23. Ihrig, P. J.; Dobbins, J. T. Jr., 1986 Winter Conference on Plasma Spectrochemistry Abstracts, Paper No. 44.
24. Barrett, P.; Davidowski,L. J. Jr.; Penaro, K. W.; Copeland, T. R. *Anal. Chem.* **1978**, 7, 1021–1023.
25. Cooley, T. N.; Martin, D. F.; Quincel, R. H. *J. Environ. Sci. Health* **1977**, *A12*, 17–18.
26. Andok, K.; Saitoh, Y.; Takatani, A.; Takahashi, F.; Tazuya, Y.; Tsunajima, K.; Motoki, C.; Yasuoka, K.; Tyamaji, Y.; Natsuoka, C. *Kenkyu Kijo-Tokushima Bunri Daigaku* **1982**, *25*, 125.
27. Bernas, B. *Anal. Chem.* **1968**, *40*, 1682.

RECEIVED for review October 5, 1987. ACCEPTED revised manuscript January 27, 1988.

CHAPTER 5

Applications of Microwave Digestion in the Pharmaceutical Industry

Sador S. Black, Judith M. Babo, and Patricia A. Stear

"Technology means the systematic application of scientific or other organized knowledge to practical tasks".
John Kenneth Galbraith

A microwave system is used to digest pharmaceutical formulations and blood samples in preparation for trace-metal analysis. Open vessels are used in the microwave oven to facilitate sample transfer. The data obtained from the microwave system are compared to those from a micro-Kjeldahl system that had previously been in use. The analyses indicate that the two systems produce comparable results but the microwave system requires less time for digestion and less supervision.

CURRENT ATOMIC SPECTROMETRIC INSTRUMENTATION allows elemental quantitation in the parts-per-million to parts-per-trillion range. Instruments such as the inductively coupled plasma atomic emission spectrometer (ICP–AES), the flame atomic absorption spectrometer (FAAS), and the graphite furnace atomic absorption spectrometer (GFAAS) require sample preparation to obtain solutions compatible with sample introduction methods. When samples are not soluble in aqueous or acidic media, sample dissolution can become the most time-consuming step in trace-metal analyses. This condition is especially true for biological samples or large organic molecules. Organic solvents can be used in sample preparation, but not all organic solvents are compatible with the instrumentation being used, and in many instances the organic matrix of a sample must be destroyed in order to determine the metal content.

Wet ashing with strong acids is a common method for oxidizing organic matter (*1–4*) and frequently uses an open vessel on a hot plate. The disadvantages of this method are the time and supervision required and the

1450–6/88/0079$06.00/0

possibility of atmospheric contamination. The need for a fast and precise dissolution technique has led to the use of microwave energy as the source of heat in digestion processes (5–11).

Our laboratory handles a variety of samples that require digestion before trace-metal analyses. These include blood, multivitamin tablets, and aztreonam (an acid-insoluble antibiotic). Digested blood is analyzed for gadolinium, which is used in the preparation of imaging agents. During clinical studies, the amount of gadolinium present in blood must be quantitated. Cupric sulfate is used in the synthesis of aztreonam, and the residual copper in the final product must be quantitated after digestion. Chromium, an essential trace element in nutrition (12), is commonly added in small quantities to multivitamin tablets. Because chromium may be present as a trace contaminant in various raw materials, a method for the quantitation of chromium in vitamins was needed to ensure that the final chromium content was within the product specifications.

The samples are routinely digested and analyzed by using an open vessel on a micro-Kjeldahl heater and the appropriate analytical instrument. The digestions of the samples require up to 90 min. To decrease the time required for sample preparation, an open-vessel microwave digestion method has been developed for each of the sample types. The microwave digestions were compared with the open-vessel digestion on the micro-Kjeldahl heater. The results from both methods were compared and demonstrate the validity of the microwave digestion technique for atomic spectrometry. Excellent agreement between the two digestion methods was obtained.

Comparative Evaluation

For each of the sample types (blood, aztreonam, and vitamin tablets) six samples were prepared and digested with the micro-Kjeldahl digester and six were digested with the microwave system. For the preparation of blood and aztreonam samples, the microwave digestion required slightly larger amounts of acid. This was to prevent charring of the sample that can occur easily in the microwave oven when open vessels are used.

Digestion Systems

A laboratory microwave system MDS-81D (CEM Corporation, Indian Trail, NC) was used for all microwave digestions. The traditional open-vessel digestions were carried out with a micro-Kjeldahl digestion rack (Labconco 60300). The rack accommodates six 30-mL Kjeldahl flasks that are placed on individual 200-W electric heaters. Each heater station has a heat control with high and low settings, as well as a variable input from 20 to 100% of capacity.

Digestion Flasks

Kjeldahl flasks (30-mL) were used as dissolution containers. The MDS-81D was equipped with Teflon [poly(tetrafluoroethylene)] sample vessels with caps that were designed for closed-vessel microwave dissolution. Two concerns when using the Teflon vessels were the quantitative transfer of the digest from the vessel and the possible loss of sample if the pressure in the vessel forced the relief valve open. Using an open vessel similar to that used with the micro-Kjeldahl digester seemed a reasonable solution to these concerns.

Initial studies showed that the high pressures obtained with the closed vessels were not necessary; microwave digestion at atmospheric pressure in the open Kjeldahl flasks was sufficient. The acid fumes generated during the digestion were removed by the exhaust system of the microwave oven and vented into the laboratory hood system. For 2 years, no corrosion of the oven cavity or exhaust system has been noted. No loss of copper, chromium, or gadolinium occurred from the open vessel during digestion, although loss of more volatile elements (e.g. silicon) has been observed while developing microwave digestion methods for other compounds.

The Kjeldahl flasks were held in position in the microwave with a notched, circular piece of polyethylene (Figure 5.1) placed in the collection vessel in the center of the single-tiered turntable. The collection vessel is used to contain any acid vapors that may escape from sealed sample vessels. The size of the openings for the vessels on the turntable required that 100-mL Kjeldahl flasks be used. The size of the flasks and the position in which they were held allowed only three samples to be digested at a time. Acid-washed glass beads were added to all flasks to prevent bumping and possible sample loss. All dissolution containers were acid washed for 4 h (minimum) and repeatedly rinsed with deionized water.

Reagents

Deionized water (18-MΩ) from a filtration system (Millipore) was used for all solutions. Both Baker "Instra-analyzed" nitric acid and Mallinckrodt AR Select nitric acid were used interchangeably and showed no discernible difference in the degree of trace metal contamination. Baker "Instra-analyzed" sulfuric acid was also used. For analyses that were performed with GFAAS, double subboiling-distilled nitric acid was used (Seastar Chemicals, Seattle, WA). Standard solutions (1000 μg/mL) of Cu, Cr (Alfa Products) and Y and Gd (Aldrich) were used as received. Tyloxapol (Triton X-100, Fisher) was used in the gadolinium analysis as a matrix modifier. All working standards were made by diluting the appropriate stock solution with 0.3 M nitric acid.

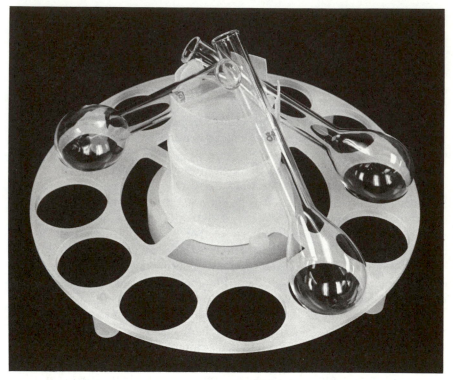

Figure 5.1. Apparatus for holding Kjeldahl flasks in the microwave oven.

Digestion Methods

Gadolinium in Dog Blood

Analysis for gadolinium in dog blood is illustrated in Figure 5.2.

Micro-Kjeldahl Digestion. Blood (1 mL) was obtained from a dog with no exposure to Gd, and was pipetted into a 30-mL Kjeldahl flask. Concentrated nitric acid (4 mL) was added to the flask. The sample was then spiked with 0.5 mL of a 100-µg/mL Gd solution. Six of these samples were digested together on the micro-Kjeldahl digester on the low setting until a reaction was detected. The heat was increased slowly, and the samples were allowed to digest until the solutions turned pale yellow. The samples were removed from the heat when their volume was approximately 0.5 mL. The time required for the digestion of the six samples was approximately 75 min. The samples were allowed to cool to room temperature before dilution.

Microwave Digestion. Dog blood (1 mL) was pipetted into a 100-mL Kjeldahl flask. Nitric acid (5 mL) was added to the flask. The samples were spiked with 0.5 mL of a 100-µg/mL Gd standard. Three samples were

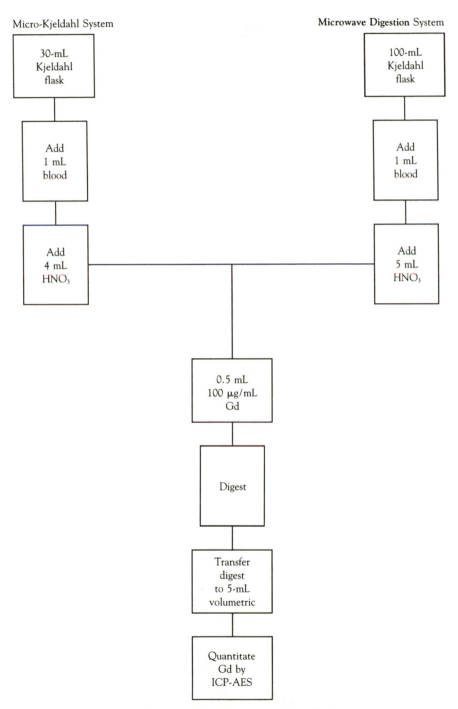

Figure 5.2. Flow chart for preparation of dog blood samples.

Table 5-1. Microwave Digestion Parameters
for Dog Blood

Step	Time (min)	Power (%)
1	2	30
2	2	50
3	3	70
4	2	80

NOTE: Total time for six samples was 18 min. Three samples per cycle.

digested simultaneously by using the time and power steps in Table 5-1. The high-power steps at the end of the cycle reduced the volume of the digest to approximately 0.5 mL. Frequently, an extra 30–60 s at 80% power was needed to obtain the correct volume. The time required to process six samples was 18 min. The samples were allowed to cool to room temperature before dilution.

Sample Preparation. Both sets of samples were treated in the same manner after digestion (Figure 5.2). Each sample was quantitatively transferred to a 5-mL volumetric flask. A 100-μL aliquot of the the 1000-μg/mL yttrium standard and 50 μL of an 80% solution (v/v) of Triton X-100 was added to each flask. The samples were then diluted to volume with deionized water.

The gadolinium content of each sample was quantitated with a Perkin–Elmer 6000 ICP–AES. The operating conditions for the ICP were as follows:

- plasma gas flow (argon), 11 L/min
- auxiliary gas flow (argon), 0.3 L/min
- nebulizer pressure (argon), 26 psi
- forward power, 1.25 kW
- reflected power, < 5 W
- viewing height above load coil, 16 mm
- wavelengths: Gd, 335.05 nm; Y, 320.33 nm

The ICP yielded the best detection limit for gadolinium with the emission monitored at 355.05 nm. Yttrium, which has a nearby emission line, was used as an internal standard to correct for matrix interferences in the sample transport. Triton-X was added to samples and standards to minimize these interferences. The results of the analysis are shown in Table 5-2.

Table 5-2. Gadolinium Concentration in Dog Blood after Digestion

Sample	Microwave Digestion[a]	S.D.	Micro-Kjeldahl[a]	S.D.
1	10.09	0.21	9.81	0.11
2	10.20	0.05	9.60	0.17
3	9.16	0.28	9.77	0.15
4	10.06	0.18	9.82	0.05
5	10.07	0.10	9.97	0.07
6	10.32	0.33	10.16	0.26
Mean	9.98		9.86	
S.D.	0.42		0.19	
C.V.	4.2%		1.9%	

NOTE: Each result is the average of three readings with an integration time of 2 s for each reading. S.D. is standard deviation; C.V. is coefficient of variation.
[a]The values given are micrograms of Gd per mL of sample solution.

Copper in Aztreonam

Analysis for copper in aztreonam is illustrated in Figure 5.3.

Micro-Kjeldahl Digestion.   Approximately 500 mg of aztreonam was accurately weighed and transferred to a 30-mL Kjeldahl flask. Nitric acid (4 mL) and sulfuric acid (0.3 mL) was added to each flask. The sample was heated on the digester at the lowest possible setting. After the initial vigorous reaction subsided, the heat was increased until the solution boiled gently. The solution was concentrated to a final volume of 1 mL. The digestion required approximately 90 min for six samples. The solutions were allowed to cool to room temperature before further sample treatment.

Microwave Digestion.   Approximately 500 mg of aztreonam was accurately weighed and transferred to a 100-mL Kjeldahl flask. Nitric acid (5 mL) was added to the flask. The cycle used in the digestion is shown in Table 5-3. The final step was adjusted to obtain the final volume of 1 mL.

Table 5-3. Microwave Digestion Parameters for Aztreonam

Step	Time (min)	Power (%)
1	2	40
2	5	30
3	2	50
4	2	60
5	1	70

Total time for six samples was 24 min. Three samples per cycle.

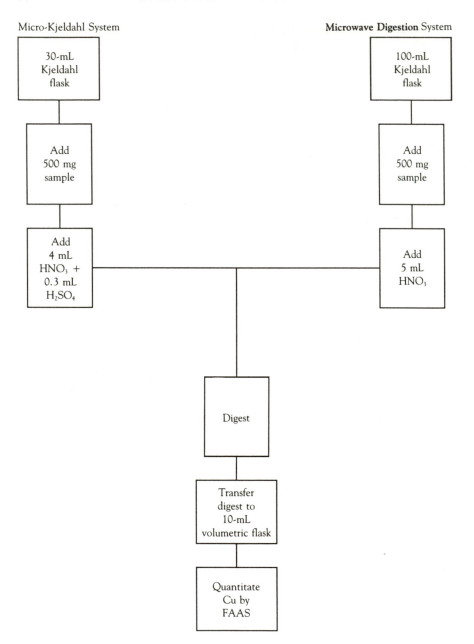

Figure 5.3. Flow chart for preparation of aztreonam samples.

The time required for six samples was 24 min. The samples were allowed to cool before any further sample treatment.

Sample Preparation.   After digestion, the samples were quantitatively transferred to 10-mL volumetric flasks and diluted to volume with deionized water. All samples were analyzed for copper by FAAS with a Varian SpectrAA-40P. The instrument operating conditions were as follows:

- wavelength, 324.8 nm
- slit width, 0.5 nm
- height above burner, 8 mm
- air–acetylene flame, fuel lean
- copper hollow-cathode lamp, 20 mA

The results of the analysis are shown in Table 5-4.

Chromium in Vitamin Tablets

Analysis for chromium in vitamin tablets is illustrated in Figure 5.4.

Micro-Kjeldahl Digestion.   Two vitamin tablets of known weight (approximately 1.2 g each) were transferred to each 30-mL Kjeldahl flasks. Nitric acid (6 mL) was added to the flasks. The samples were then placed on the microdigester, and the heaters were turned to the low setting. After the initial reaction, the heat was turned up gradually until the sample volume was reduced to approximately 1 mL. The time required for the digestion of six samples was 90 min. The samples were allowed to cool before any further sample treatment.

Table 5-4. Copper Concentration in Aztreonam

Sample	Microwave Digestion (μg of Cu/g)	Micro-Kjeldahl (μg of Cu/g)
1	9.0	8.8
2	8.7	9.1
3	9.0	8.9
4	9.2	8.8
5	9.2	8.8
6	9.3	8.8
Mean	9.1	8.9
S.D.	0.22	0.12
C.V.	2.4%	1.4%

NOTE: Data from Varian SpectrAA-40P are the average of three readings with an integration time of 2 s for each reading. S.D. is standard deviation; C.V. is coefficient of variation.

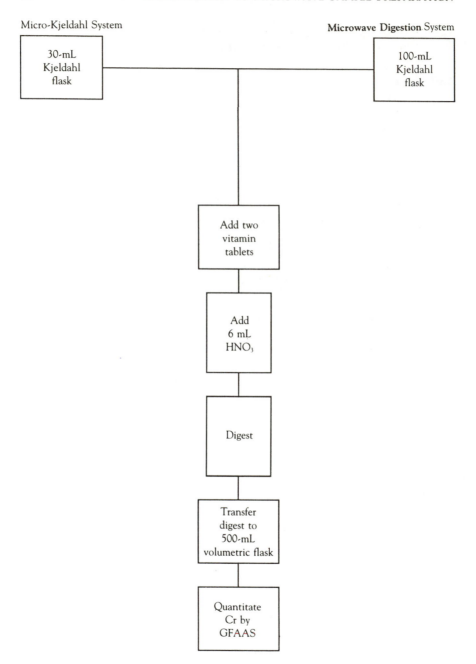

Figure 5.4. Flow chart for preparation of vitamin samples.

Table 5-5. Microwave Digestion Parameters for Vitamin Tablets

Step	Time (min)	Power (%)
1	2	20
2	3	40
3	1	60
4	1	40

Total time for six samples was 14 min. Three samples per cycle.

Microwave Digestion.   Two vitamin tablets of known weight were transferred to each 100-mL Kjeldahl flask. Nitric acid (6 mL) was added to the flasks. The samples were digested in the microwave system in sets of three by using the program found in Table 5-5. The time required to prepare six samples was 14 min.

Sample Preparation.   The contents of each flask were transferred to separate 500-mL volumetric flasks, and 10 mL of nitric acid was added to bring the final acid concentration to approximately 0.3 M. Samples were diluted to volume with deionized water, filtered, and analyzed for chromium by GFAAS using a Perkin–Elmer HGA 500 equipped with an AS-40 autosampler. The operating conditions for the instrument were as follows:

- wavelength, 357.9 nm
- spectral bandwidth, 0.7 nm
- background correction, tungsten–halogen lamp
- chromium hollow cathode lamp, 25 mA
- sample aliquot, 20 μL
- matrix modifier, 5 μL
- total sample volume, 25 μL
- pyrolytic graphite tube with platform

The heating program is shown in Table 5-6.

Initially, the vitamin digest was transferred to a 100-mL volumetric flask, in an attempt to increase the Cr concentration in the sample solution. However, the complicated vitamin matrix caused spectral interferences in ICP analysis and the chromium concentration was still below the optimum working range for FAAS. Therefore, the graphite furnace was necessary to quantitate the chromium present in the vitamin tablets. The results for the samples are shown in Table 5-7.

Table 5-6. Heating Program for Perkin-Elmer HGA 500 Graphite Furnace (Chromium Analysis—Multivitamins)

Step	Temperature (°C)	Ramp Time (s)	Hold Time (s)	Internal Gas Flow (mL/min)
1	120	10	20	300
2	1400	10	30	300
3	20	5	10	300
4	2550	0	5	0
5	2700	1	4	300
6	20	10	10	300

Results and Discussion

The microwave digestion system was incorporated into our routine sample preparation to decrease both the time required for sample preparation and the need for supervision by the analyst. Use of the microwave system reduced the sample dissolution time by 57 min for blood, 66 min for aztreonam, and 76 min for the vitamin tablets. Less supervision was required because the system can be programmed and no further attention will be needed until the microwave oven signals that the cycle is completed. On the other hand, almost constant supervision is required for the micro-Kjeldahl digester because the progress of the digestion must be monitored and heating adjustments made.

In addition, there are several other advantages to the microwave system. The heating elements on the micro-Kjeldahl digester do not heat uniformly, and continuous adjustments must be made to maintain the same rate of digestion for all samples. The microwave digester uses a turntable to deliver the same intensity of microwave energy to all samples. Also, the heating

Table 5-7. Chromium Concentration in Vitamin Tablets

Sample	Microwave Digestion (μg of Cr/g)	Micro-Kjeldahl (μg of Cr/g)
1	5.6	5.8
2	5.3	5.2
3	4.8	5.4
4	5.2	5.0
5	4.8	5.1
6	5.3	5.4
Mean	5.2	5.3
S.D.	0.31	0.29
C.V.	6.1%	5.4%

NOTE: Mean represents the average of three values. S.D. is standard deviation; C.V. is coefficient of variation.

cycles for the microwave digester are very reproducible because the power levels and times are controlled electronically.

One disadvantage of the microwave digestion system that also applies to the more conventional micro-Kjeldahl digester is the empirical derivation of the heating cycles for a given sample. Although Kingston and Jassie (*13*) developed temperature profiles for acid digestions in a microwave system, the data applied to closed-vessel systems.

In our studies, choice of the digesting acids and the volume of acids used were based on the micro-Kjeldahl digestions. This approach was designed to decrease the time required for the digestions that were currently in use. Modifications were necessary in two of the three methods. When using an open vessel in the microwave system, a minimum of 5 mL of acid was required to prevent the charring of the sample before digestion was complete. Charring rarely occurred with the micro-Kjeldahl system because the digestion proceeded very slowly.

In the digestion of dog blood, 5 mL of nitric acid was used in the microwave digester as opposed to 4 mL with the micro-Kjeldahl digester. Sulfuric acid was used in the digestion of aztreonam with the micro-Kjeldahl heater, but the use of sulfuric acid is not recommended in the microwave digester because of the high boiling point of the acid. A larger volume of nitric acid was used for the microwave system. No changes were necessary for the digestion of the vitamin tablets. Several trials were necessary to determine the optimum microwave digestion conditions of power and time for each sample; however, once these conditions were determined, any number of sample sets could then be digested consecutively and reproducibly.

The analytical results for each sample are shown in Tables 5-2, 5-4, and 5-7. The tables show that the microwave digestion system can be used to produce analyses comparable to the micro-Kjeldahl digestion. The mean values reported for the two digestion techniques for each sample are not statistically different and there is no significant difference in the standard deviation of the analyses. The excellent results that have been observed in our laboratory have prompted the use of the microwave system for all current digestions. In addition, microwave digestion is now the method of choice when developing sample dissolution procedures for new assays.

Literature Cited

1. Gorsuch, T. T.; *The Destruction of Organic Matter*; Pergamon: Oxford, England, 1970; pp 19–26.
2. Brock, R. *Decomposition Methods in Analytical Chemistry*; translated and revised by Marr, I. L.; International Textbook: London, 1979.
3. Krishnamurty, K. V.; Shpirt, E.; Reddy, M. *At. Absorption Newsletter* **1976,** *15,* 68–70.
4. Reamer, D.; Veillon, C. *Anal. Chem.* **1983,** *55,* 1605–1606.

5. Fischer, L. *Anal. Chem.* **1986,** *58,* 261–263.
6. Fernando, L; Heavner, W.; Gabrielli, C. *Anal. Chem.* **1986,** *58,* 511–512.
7. Lamothe, P.; Fries, T.; Consul, J. *Anal. Chem.* **1986,** *58,* 1881–1886.
8. Abu-Samra, A.; Morris, J.; Koirtyohann, S. *Anal. Chem.* **1975,** *47,* 1475–1477.
9. Barrett, P.; Davidowski, L.; Penaro, K.; Copeland, T. *Anal. Chem.* **1978,** *50,* 1021–1023.
10. White, R.; Douthit, G. *J. Assoc. Off. Anal. Chem.* **1985,** *68,* 766–769.
11. Nadkarni, R. *Anal. Chem.* **1984,** *56,* 2233–2237.
12. Czarnecki, S.; Kritchevsky, D. In *Nutrition and the Adult: Micronutrients;* Alfin-Slater, R.; Kritchevsky, D., Eds.; Plenum: New York, 1980; pp 319–350.
13. Kingston, H.; Jassie, L. *Anal. Chem.* **1986,** *58,* 2534–2541.

RECEIVED for review July 24, 1987. ACCEPTED revised manuscript May 9, 1988.

CHAPTER 6

Monitoring and Predicting Parameters in Microwave Dissolution

H. M. Kingston and L. B. Jassie

"If I have seen further it is by standing upon the shoulders of Giants".

Sir Isaac Newton

Procedures are described for the real-time measurement of temperature and pressure during closed-vessel microwave sample decomposition. Pressure and temperature profiles of biological Standard Reference Materials and solitary as well as mixed acids are given to illustrate unique advantages that are available with the closed-vessel technique. A set of equations that per.nits prediction of target temperatures and times is derived from the fundamental heat capacity relationship for absorptive materials. From a series of fundamental measurements, original equations are introduced that permit the power consumption of common mineral acids to be calculated. This method is proposed as a model to approximate the thermal behavior of reagents intended for microwave use. The fundamental degradation patterns of biological matrices are presented for model compounds.

UNDERSTANDING OF THE PROCESSES THAT OCCUR during microwave-heated acid decompositions has been aided by the ability to measure temperature and pressure in situ. Like water, acids are dielectric materials and absorb electromagnetic energy from the microwave field. Different acids absorb different quantities of microwave energy, and a single acid absorbs different amounts of energy depending on its concentration and on the total mass present. Decomposition conditions are experimentally repeatable. If the length of exposure and available power are known, both the temperature and pressure can be predicted. Equations are derived that predict the temperature of a particular quantity of acid at a given applied power.

Monitoring conditions in closed vessels is more important than in open vessels, which are at atmospheric pressure, because in open-beaker decom-

1450–6/88/0093$15.30/0

positions, the temperature is limited to the boiling point of the acid or the azeotropic mixture of acids. This maximum temperature is maintained until all the acid has evaporated. If many acids are present, the most volatile acid will boil away first, followed by the next most volatile, and so on, unless an azeotropic mixture is formed. In open-vessel decompositions, reactions proceed at the boiling point of the mixture. When closed vessels are used, however, the solution temperature is not limited by the boiling point of the acid at atmospheric pressure, and for many acids the boiling point can be raised significantly higher than in open-vessel procedures. The limiting parameters in closed-vessel decompositions are the temperature and pressure that the vessel can safely contain. Once the safety limits of a vessel are established, temperature and pressure can be monitored to maintain the reaction within these limits.

Closed-system acid decompositions have many advantages over atmospheric decompositions. Nitric and hydrofluoric acids can be used to to decompose materials that do not react at atmospheric pressures and temperatures. The elevated temperatures produce significant increases in oxidation potential and form intermediates, such as free radicals, that facilitate the chemical attack of compounds (1). In addition to increasing reaction rates, the closed system minimizes contamination from laboratory air and reduces the amount of acid necessary in the decomposition, thus reducing the analytical blank associated with the sample preparation. Volatile trace elements that would normally be lost in open systems are retained in closed systems.

Temperature and Pressure Measurement in Closed Vessels

Monitoring Parameters

The physical requirements of the sample containers used for this application are so demanding that only a few materials are suitable for vessel construction. Although polycarbonate and other materials have been used as containers for closed-vessel microwave acid digestions (2), pressure limitations, acid resistance, and frequent vessel failures have restricted their use. The container material must be transparent to microwave radiation and inert to mineral acids at temperatures > 200 °C. The safe operating pressure of the vessel establishes the upper limits for pressure and temperature that can be used in the digestion. Chemical resistance, tensile strength at high temperatures, and microwave transparency of Teflon PFA (perfluoroalkoxy) make it the most appropriate material. This polymer's superior chemical characteristics and improved mechanical features, including innovative thread design and increased wall thickness, allow it to be safely used at elevated pressure and temperature.

Teflon PFA digestion vessels, specifically designed for microwave use under these conditions, are manufactured by the Savillex Corporation; they are satisfactory for monitoring temperature and pressure during closed-container operation. The Teflon PFA vessel is manufactured with a variety of ports that are molded as integral parts of the cap. The cap configuration with two 1/8-in. ports is used for the monitored vessels. Caps without ports are used on unmonitored containers. Retrofitting a vessel to provide ports for measurement probes can compromise the vessel's integrity. Other vessels (a high-pressure microwave digestion bomb) have recently been introduced for elevated temperature and pressure microwave acid digestions, but they do not permit internal measurements.

The development of special sample vessels for use in analytical chemistry has aided the field of microwave acid decomposition by overcoming specific laboratory problems. The introduction of chemically inert Teflon PFA polymer in the construction of closed vessels has permitted advances in microwave digestion procedures not previously possible.

Moderate-Pressure Closed Vessel

Because most microwave acid decompositions are performed in closed containers, the design and construction of these vessels is very important. The specially engineered all-Teflon PFA pressure vessel is available equipped with a safety pressure-relief disk to prevent overpressurization. Figure 6.1 shows an expanded view of the vessel and pressure-relief disk. The internal pressure of this vessel will not exceed 120 ± 10 psi. Pressures greater than the rated "use pressure" of the vessel will develop for nitric acid temperatures > 185 °C. The pressure-relief disk forms a seal against the inside of the cap around the pressure-relief port. When the gas pressure exceeds 120 psi, the top of the cap acts as a diaphragm, flexing away from the disk seal and allowing gas to escape through the venting port relieving the pressure > 120 psi. A cross-sectional diagram of this relief mechanism is shown in Figure 6.2. These vessels are used routinely.

If the internal pressure does not exceed 120 ± 10 psi, the vessel remains closed and volatile components do not escape. A convenient way to verify venting is to weigh the vessel containing the sample and acid before and after decomposition. Teflon is a semipermeable membrane that allows the passage of certain gases. Small quantities of water and mineral acids such as nitric, hydrochloric, and hydrofluoric acids will diffuse into the walls and escape during decomposition. Metal cations and most anions remain in solution and do not penetrate the Teflon PFA. Exceptions to this rule of thumb are elements that can pass through Teflon PFA such as metallic mercury, osmium tetroxide, and the hydrogen halides.

If there is a great deal of agitation inside the vessel during decomposition, small amounts of sample may be expelled during venting as aerosols

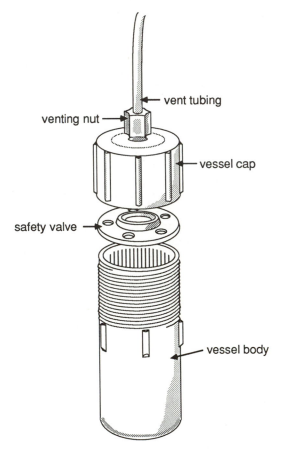

Figure 6.1. Moderate-pressure Teflon PFA digestion vessel with safety relief disk.

or condensed liquid. Analytes may be lost as gas escapes from the vessel. A venting tube may be fitted with a trap or condenser to prevent such losses. Different vessel sizes and configurations are available for a variety of applications. For example, some vessels come with double-ported caps that allow both the temperature and pressure to be monitored simultaneously. This two-port configuration has been used as a trap when used with another decomposition vessel.

High-Pressure Microwave Vessels

Because the temperature in closed-vessel acid decomposition is frequently limited by the pressure the container can safely withstand, new and stronger microwave-transparent vessels are being designed and constructed. One type

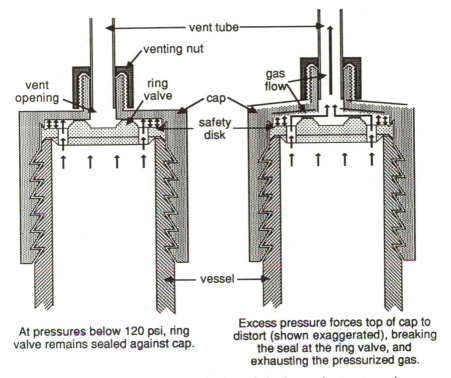

At pressures below 120 psi, ring
valve remains sealed against cap.

Excess pressure forces top of cap to
distort (shown exaggerated), breaking
the seal at the ring valve, and
exhausting the pressurized gas.

*Figure 6.2. Cross-sectional view of safety relief valve mechanism in moderate-
pressure Teflon PFA digestion vessel.*

of high-pressure device (Parr Instrument Co.) resembles the steel-jacketed
bombs used in conventional convection ovens. An inert Teflon TFE liner
and a strong, rigid retainer surround the core vessel and provide its structural
support. A polymeric retainer material that is transparent to microwave
radiation is used to provide this rigidity.

The new microwave bomb design incorporates replaceable Teflon
O-rings in the liner cap that seal against a narrow rim on the exterior of
the liner and its cap when the retaining jacket is screwed into place. When
overpressurization occurs, this O-ring is distorted (and may rupture). The
gas pressure escapes by compressing a puck-shaped disk above the liner cap
(Figure 6.3). Gas or vapor can escape through four outlets in the retainer
cap positioned at 90° intervals around its circumference. If this occurs, the
sample is compromised.

The vessel has been designed for a maximum working pressure of 80
atm, approximately 10 times the pressure that can be sustained in the unsupported Teflon PFA vessel. These high internal pressures permit increased
temperatures for the acids and acid mixtures that would otherwise over-
pressurize the unsupported PFA vessels. These temperatures cannot be sus-

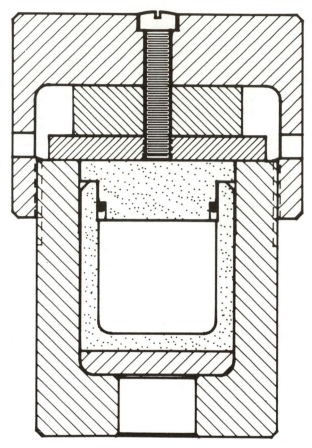

Figure 6.3. High-pressure vessel for microwave dissolution (Parr Instrument Co.).

tained for lengthy periods, because if the temperature of the outer polymeric retaining vessel is allowed to exceed 50 °C, the structural integrity necessary to maintain the seal and high pressure may be compromised. Because pressure and temperature conditions in the vessel cannot yet be determined, calculated parameters cannot be confirmed.

Because of the thickness of the inner Teflon core and light-weight but large outer jacket, the contents of the sample cup are well-insulated. This insulation significantly reduces heat loss by conductive heat transfer from the sample to the vessel walls. Conditions that would have taken more than 1 h to reach in a conventional oven with the steel-jacketed bomb can now be achieved in less than 1 min. To prevent blowout and subsequent sample loss, it may be best to approach conditions of extreme pressure slowly by using partial power and multiple-step heating.

Temperature Measurement

Temperature measurement during operation is complicated by the high-energy microwave fields in the cavity. Conventional temperature measurement devices cannot be used in this environment. Three measurement systems have been developed for use in microwave environments. One is a shielded thermistor like those found in conventional home microwave appliances. This temperature probe is not suitable for use in microwave acid digestion for several reasons. Temperatures during acid digestions range from ambient to just over 400 °C with sulfuric acid. Conventional microwave appliance thermistor probes have an upper limit of just over 100 °C. Shielding of the thermistor is accomplished by encasing it in a stainless steel tube electrically connected to a copper braid grounded at the oven wall. Successful operation in the microwave field requires that the end of the probe be placed inside a microwave absorber where it is essentially isolated from the field. A microwave absorber, such as water or hydrated tissue, totally consumes the 2450-MHz field within a distance of approximately 2.5 cm. This conventional shielding has proven unsuitable for most laboratory applications, because probes exposed to the microwave field without ample protection by a dielectric absorber can accumulate a charge sufficient to produce electrical arcing.

Temperature measurement during acid decomposition in the microwave field has been successfully accomplished with thermocouples. For most acid mixtures in Teflon vessels, a copper–constantan thermocouple works well. The thermocouple can be chosen to provide accurate measurement in the temperature range desired. Although the type of thermocouple does not influence its ability to function in a microwave field, all such metal probes must be completely shielded to be used in this environment. Configurations used for sample decomposition frequently require the measurement of small quantities of acid and sample, typically 3–10 mL. These small volumes leave the probe completely exposed to the full intensity of the microwave field. Proper shielding and electrical grounding of the shielding are essential for accurate temperature measurement and reliable probe operation under these conditions. Both the probe and connecting braid must be grounded at the wall of the unit to dissipate the electrical charge that accumulates on the probe and shielding. The path to ground requires the lowest resistance achievable. Gold plating both the tip and the braid of the shielding is an effective way to reduce the resistance. The construction of a shielded thermocouple sensor, with proper dimensions for use in a microwave cavity, has been described in detail previously (3, 4).

The use of fiber optic devices (Luxtron Corp., Model 750) for temperature measurement in microwave fields is a relatively recent development. Unlike metal-shielded thermistors and thermocouples, the fiber optic device does not interact with microwave energy and is transparent in the field.

Although the current instrumentation is relatively expensive, it is reliable and accurate (5–7). The use of this instrumentation in closed-vessel microwave acid digestion has specific equipment configuration requirements (8).

Because the Teflon PFA probe coating is permeable to acid vapor, the fiber optic probe cannot be placed directly into the acid. The sensing phosphor at the tip of the probe and the Kevlar (polyaramide fiber) protective covering are attacked by strong oxidizing acids and should be protected by inserting the first 12 cm (5 in.) of the probe into a 1/8-in.-thick-walled Teflon PFA tube that is sealed at one end. This sleeve can then be inserted

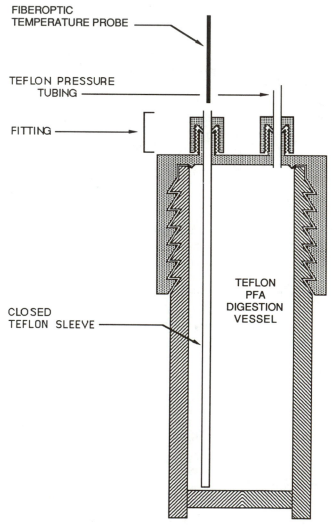

Figure 6.4. Teflon PFA digestion vessel with fiber optic temperature probe and pressure line.

into the digestion vessel cap port and sealed to hold pressure with ferruled Teflon nuts. The pressure and torque of sealing the fiber optic probe directly could cause the glass fiber to break at this point; direct pressure on the probe itself should be avoided to protect the fragile glass filament. Figure 6.4 illustrates a configuration that protects the fragile probe from the effects of high pressure and chemical attack by nitric, phosphoric, hydrochloric, and dilute hydrofluoric acid (and combinations of these acids), and yet provides reliable temperature measurements. Problems encountered with this measurement system are related to the durability of the fiber optic probes; the probes are easily broken and chemically attacked by long exposure to concentrated hydrofluoric acid vapor.

The Teflon PFA sleeve does not protect the probe from repeated use in concentrated solutions of hydrofluoric acid. Degradation of the probe and subsequent failure have been experienced after 8–10 digestions of 10–15 min with mixtures containing 50% hydrofluoric acid. Electron microscopy indicated that chemical attack of the glass fiber was the cause of probe failure (9). To prevent accumulation of acid vapor in the permeable Teflon coating on the surface of the probe, the probe should be removed from the Teflon sleeve and rinsed with water or a dilute solution of ammonium hydroxide. Rinsing will prolong the useful life of the probe. Sharp bending of the probe should be avoided to prevent breaking the glass fiber.

A number of experimental probe configurations using optical links and remote phosphor scanning are being investigated. A remote temperature measurement technique using phosphors on the outside of glass digestion vessels has been used with high-temperature sulfuric acid decompositions (10). This technique eliminates the necessity of placing the probe in the digestion vessel but introduces a bias into the reading because of the cooling effect of the air on the outer surface of the container. In this configuration, the temperature measured is that of the glass surface and not of the acid and sample inside. This remote technique is not suitable for use with the Teflon vessels without modification because of the tremendous insulating capability of the Teflon.

When temperature and pressure probes are needed inside the microwave cavity, a 360° reversing turntable should be used to avoid entanglement of the sensors. The center of the carousel or top of the microwave cavity can be used to anchor the temperature and pressure probes, and to facilitate separation of the sensor lines.

Pressure Measurement

Pressure in the Teflon PFA vessel is monitored through a 1/8-in.-thick-walled Teflon PFA tube connected to one of the fittings on the cap of the container as shown in Figure 6.4. Because the hot vapor in the vessel may condense in the pressure tube and be drawn back into the vessel at the end of the heating cycle, a short section (approximately 20 cm) of the pressure tubing is cleaned and replaced each time the vessel is changed to prevent cross contamination of the samples. This tube is connected to a Teflon PFA "T" or union mounted near the center of the turntable; tubing from the other end of the union continues out of the microwave cavity through a 1/8-in. hole in the wall and through a flexible wavelength attenuator cutoff attached directly over the opening to prevent the escape of microwave energy. A pressure-sensitive adjustable valve is placed in the pressure line just before a pressure transducer. Figure 6.5 illustrates the arrangement of both the temperature and the pressure-monitoring equipment in conjunction with the microwave system (modified CEM Corp. MDS-81) as previously described (4). The functional range of both the transducer and the valve should be matched to the pressure range and vessel used. For Teflon vessels, the functional range of the relief valve should be from 5 to 10 atm (1 atm ≃ 1 × 10⁵ Pa); thus, the transducer should cover a pressure range from ambient atmosphere to 17 atm.

Whereas the vessel and pressure tubing are ordinarily Teflon, the pressure-measurement device and safety valve are stainless steel. The steel components must be isolated from the contents of the vessel and from the pressure line to prevent contamination of the samples and degradation of the metallic

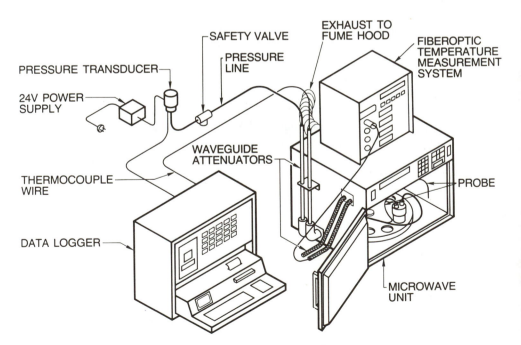

Figure 6.5. Microwave digestion unit, pressure- and temperature-sensing equipment, and data acquisition system.

components. A trap of distilled water can be used to isolate the transducer. Water is noncompressible and transmits the pressure without allowing the vapor to contact the metal portions of the remote pressure-measurement system. If water is used, it must be changed frequently, because it will accumulate dissolved acid vapor from the vessel.

Temperature and Pressure Data Acquisition

Temperature and pressure data can be obtained with a data acquisition system. This task has become easier and more routine in chemistry laboratories with recent advances in microcomputers. Pressure and temperature data can be recorded directly using an analog-to-digital converter card in a laboratory computer system. The fiber optic instrument can be equipped with an RS232 serial output that can be coupled directly to the computer. Small temperature measurement devices that allow direct digital reading and recording or displaying of the data are available for use with thermocouples. Algorithms have been written to plot temperature and pressure vs. time to document the actual conditions during the digestion. Plotting these parameters vs. time or one another is useful for understanding the interactions of the acid with the microwave energy and the acid decomposition of the sample. Temperature and pressure data are used in the development of decomposition schemes and in the quality control of routine dissolutions.

Data acquisition equipment can be assembled in many different ways to record and display the temperature and pressure data from the microwave digestion. The configuration used depends largely on the sensors used in monitoring and the needs of the analyst. If there is no need to record data, direct-reading devices that provide digital display of both temperature and pressure are available.

Calibrating Instruments, Measuring Power, and Predicting Conditions

Thermodynamic Basis of Measurement

Equations for power measurement in a microwave system are derived from elementary theory of the heat capacity of a given mass at constant pressure. Heat capacity, C_p, is that quantity of heat required to raise the temperature of a given mass by 1°. The energy absorbed produces a rise in temperature, ΔT. If a quantity of energy is delivered for a unit of time, then P, the power absorbed by a substance (power density) in the microwave cavity may be expressed in the following relationship

$$P_{absorbed} = \frac{KC_p m \Delta T}{t} \qquad (6.1)$$

where P is the apparent power absorbed by the sample in watts, (1 W = 1 J·s^{-1}); K is the conversion factor for thermochemical calories per second to watts (4.184 J·cal); C_p is the heat capacity, thermal capacity, or specific heat (cal·g^{-1}·°C^{-1}); m is the mass of the sample in grams; ΔT is T_f, the final temperature, minus T_i, the initial temperature (°C); and t is the time in seconds.

Power Measurements and Equipment Calibration

Equation 6.1 has been used, with minor modifications, to quantitatively establish the significance of local variations in tissue temperature as related to changes in the heat content of the body as a whole (11) as well as to express the absorbed power density of tissue exposed to electromagnetic radiation (12–14). On the basis of the assumption that the majority of the power delivered to the cavity is absorbed by a sufficiently large quantity of a dielectric absorber, equation 6.1 may be used to evaluate the power output of a magnetron to the cavity.

Each microwave system is a unique device; the power delivered to the sample depends not only on the power output of the magnetron but also on the tuning of the waveguide and cavity dimensions. The apparent power absorbed by water irradiated at full power should be used to calibrate all 2450-MHz microwave equipment that have power outputs between 500 and 800 W. This calibration is accomplished by measuring the temperature rise of 1 kg of water that has been exposed to electromagnetic radiation for a fixed period of time. Although several different methods have been used (4, 15–18), satisfactory measurements can be made on replicates of weighed, 1-kg samples of room-temperature distilled water in thick-walled microwave-transparent vessels. Teflon and polyethylene containers are among the most microwave-transparent and have been used successfully for these calibrations. Potential sources of error in this determination include using microwave-absorptive or reflective containers, not stirring the water before measuring, and losing heat from the vessel. The container should be circulated continuously through the field for at least 2 min at full power. Although 2 min is a compromise between short and long exposures and produces a small change

Table 6-1. Apparent Power Absorbed by 1 kg of Water

Time (s)	Power (W)
90	570 ± 1
120	574 ± 7
150	569 ± 19
180	573 ± 8
210	562 ± 9
240	565 ± 6

NOTE: All power calculations are ± 1 standard deviation; $n = 3$–5.

in temperature, 1-kg samples can be exposed for 90–240 s at full power without significant differences in apparent power absorption (Table 6-1).

In a homogeneous microwave field, 1 kg of water absorbs approximately the same amount of power in one container as it does when it is equally divided between two or five containers.

Measurement of heat transfer is critical to the calibration of microwave equipment. A variety of microwave equipment is available that delivers from 500 to nearly 800 W of power to the cavity. The actual power delivered by each magnetron must be determined so that absolute power settings can be interchanged from one microwave unit to another. Previously designed digestion schemes can be adapted to different systems when the number of watts of power available to the samples is known. A microwave unit that has its magnetron protected from reflected microwave energy can be expected to continue producing constant power over many years of operation. For example, the magnetron in the equipment used in this study is protected from reflected microwave energy by an isolator and has maintained a constant 574 ± 7 W for over 4 years of frequent operation.

Equation 6.1 can be used to calculate the power uptake of any quantity of material for which the heat capacity is known and for which initial and final temperatures can be measured (4). Heat capacity values in the literature are frequently given as the apparent molal heat capacities, (Φ_c, in calories per degree centigrade per mole). The heat capacities (calories per gram per

Table 6-2. Heat Capacity of Mineral Acids and Solutions

Acid Solutions	Concentration $(mol \cdot L^{-1})$	Heat Capacity $(cal \cdot g^{-1} \cdot C^{-1})$
Acetic (100%)	17.4	0.4947
Hydrochloric (37.2%)	12	0.5863
	6	0.7168
	1	0.9378
Hydrofluoric (49%)	28.9	0.6960
Nitric (70.4%)	15.9	0.5728
	8	0.7162
	1	0.9497
Phosphoric (85.5%)	14.8	0.4470[a]
Sulfuric (96.06%)	18	0.3499[b]
	6.7	0.6142[b]
	1.1	0.9142[b]
Sodium chloride (in water)	1	0.9339
Water	55	0.9997[c]

NOTES: All data are normalized to 25 °C.
SOURCE: Adapted from reference 19.
[a]Reference 36.
[b]Reference 37.
[c]Reference 38.

degree centigrade) have been calculated from values and equations given by Parker (*19, 20*) for commonly used acids at several concentrations and are summarized in Table 6-2. The heat capacity of an acid varies inversely with its concentration. The more dilute the acid the greater its heat capacity, and the more nearly its value approaches that of water.

Power absorption by small amounts of acids decreases proportionally as the mass in the cavity decreases (Table 6-3), because an increased portion of the incoming radiation never travels through the sample as the waves traverse the cavity (*see* Chapter 2).

Microwave power absorptions are compared in Table 6-3. The table shows that dilute acids absorb power more strongly than concentrated acids. This difference is attributable to the larger fraction of water present in dilute acids. Water is a better absorber of 2450-MHz radiation than any of the mineral acids measured. Values for the absorbed power are derived from measurements made at full power for different time intervals, so that the net temperature change is between 5 and 50 °C. Large temperature increases are accompanied by heat losses and changes in heat capacity; neither condition is considered in equation 6.1. Empirically, it can be observed that very small samples get hotter than large samples for the same power setting and exposure time. This observation seems logical, because there is less mass to heat and the microwave power density is greater. Small acid quantities actually absorb proportionally less power than large quantities because they reduce the energy content of the wave to a lesser degree. For example, at an exposure to full power (574 W) for 2 min, the temperature of 200 mL of 6 M HCl increases by 50 °C; whereas, 500 mL of 6 M HCl increases by only 35 °C.

The power absorption data for 25- to 3000-g samples of mineral acids exhibits a nonlinear relationship between mass and power. Power absorption continues to increase with sample mass until relatively large sample masses are reached. For the materials studied to date, the absorbed power rises sharply with increasing mass until it reaches approximately 500 g, it levels out, and then increases very little between 500 and 1000 g.

Table 6-3. Power Absorbed by Small Volumes of Concentrated Mineral Acids Compared with Distilled Water

Reagent and Concentration	50 mL	100 mL	200 mL
H_2O	344 ± 9	408 ± 3	468 ± 5
HNO_3 (16 M)	184 ± 2	234 ± 5	313 ± 4
HNO_3 (1 M)	212 ± 3	269 ± 6	332 ± 3
HF (29 M)	167 ± 3	238 ± 12	315 ± 15
H_2SO_4 (18M)	231 ± 6	331 ± 4	396 ± 8
HCl (12 M)	148 ± 3	173 ± 6	251 ± 2
HCl (6 M)	138 ± 3	190 ± 4	253 ± 4
HCl (1 M)	227 ± 4	287 ± 7	340 ± 5

NOTES: All power calculations are in watts $\pm$ 1 standard deviation; $n = 5$.

Because the amount of power absorbed is proportional to the quantity of acid or water present in the cavity, that power can be predicted for a known mass of acid. The calculations of absorbed power plotted against the mass of the mineral acids and water are presented in Appendix A, which immediately follows this chapter. These data are of practical value because the power absorption must be known in order to predict temperature conditions in the sealed vessels during digestion.

Predicting Conditions

In laboratories that perform a large number of routine sample dissolutions, it is not efficient to digest one sample at a time. Multiple dissolutions can bring the total acid quantity in the cavity to between 25 and 1000 g, depending on the individual sample sizes. The power absorbed for any mass of acid in the cavity can be calculated with a set of equations derived from the experimental data for the acids and water.

Two models have been used to relate the absorption of power to the mass of either acid or water. Equation 6.2 is a natural logarithm-based linear equation and equation 6.3 is a natural logarithm-based quartic model of the same data (given in Appendix A). The actual coefficients A′ and B′ (linear model) and A through E (quartic model) used in these generalized equations for the mineral acids and water are given in Appendix B. The generalized expressions are as follows:

First-order (linear) model:

$$\ln \text{(absorbed power)} = A' + B' \times \ln \text{(mass)} \qquad (6.2)$$

Fourth-order (quartic) model:

$$\ln \text{(absorbed power)} = A + B \times \ln \text{(mass)} + C \times \ln \text{(mass)}^2$$
$$+ D \times \ln \text{(mass)}^3 + E \times \ln \text{(mass)}^4 \qquad (6.3)$$

NOTE: Quartic equations are especially susceptible to rounding errors; it is very important not to round off any of the coefficients provided for the quartic equation or any intermediate terms.

The difference between these two mathematical models is the accuracy with which they predict the amount of power absorbed by a mass of acid or water. Table 6-4 shows the average percent prediction error based on 95% confidence limits for both models. The first-order model is biased and does not adequately represent the data, but it provides a safety aid and computations can be performed with hand calculators for quick estimates of power absorption. The fourth-order equation represents the data with greater accuracy and should be used for prediction of conditions. Appendix C contains graphic results for both models along with the data and the 95% confidence

**Table 6-4. The Average Percent Error (at the 95%
Confidence Limit) for the Linear and
Quartic Equations**

Reagent and Concentration	Linear	Quartic
H_2O	15.5%	8.1%
H_2SO_4 (18 M)	20.5	5.4
HCl (1 M)	7.1	3.9
HCl (6 M)	12.2	5.3
HCl (12 M)	13.7	10.4
HF (18 M)	18.4	9.5
HNO_3 (1 M)	8.4	4.8
HNO_3 (16 M)	12.6	8.1

limits for each model. These graphs show the excellent fit for the fourth-order equation and provide a comparison for the predicted accuracy using the two different models. If greater accuracy is needed for a particular mass of acid or reagent, then the temperature of the actual mass of reagent needed should be measured under identical conditions and the absorbed power calculated from equation 6.1.

The absorbed power, in watts, for a given mass, in grams, of a particular acid was used to estimate parameters for equation 6.2 and 6.3. These measurements were obtained with a microwave unit that delivered 574 W of power to the microwave cavity. To make these power absorbances relevant to another microwave unit, they must be corrected by using the calibrated value at full power for that system. If the maximum power of the calibrated microwave unit is 610 W, this value should be divided by 574 to give the correction factor between microwave units (1.06 for this example). This factor should be multiplied by the absorbed power obtained from the model for a particular acid (for this example 1.06 × 250 W = 265 W) to obtain the amount of power absorbed by that quantity of acid corrected for the 610-W microwave unit. Such corrections are possible on calibrated units because proportional power delivered to the cavity results in proportional power absorption by the quantity of reagent present. Figure 6.6 demonstrates this relationship between absorbed power and applied power.

The upper power limit of the microwave unit is dependent on the output of the magnetron and cavity tuning. For very large samples, the power absorption is constrained by the actual power delivered to the microwave cavity and by the unique conductance and dielectric relaxation time of the particular acid (*see* Chapter 2). This upper limit and the mass at which it is reached will be slightly different for each acid and for microwave equipment delivering different amounts of power to the microwave cavity. Some power absorption increase can be seen beyond 1000 g for sulfuric and nitric acids. For other acids, the power absorption curve flattens at a mass of 1000 g. Equations 6.2 and 6.3 should not be used to predict power

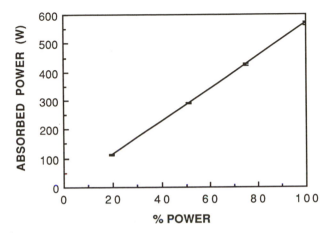

Figure 6.6. Relationship between absorbed power and applied power.

absorption for masses outside the range used to estimate parameters for these equations. These mass ranges are in Appendix A. Likewise, approximately 25 g is the lowest mass that these models can accurately predict. (This quantity of acid is the smallest that could be measured accurately during tests.)

Once a value of P has been calculated for a specific volume of acid, both the target temperature and the time that microwave power should be applied can be estimated by a simple transformation of equation 6.1. The final temperature can be estimated from

$$T_f = T_i + \frac{Pt}{KC_p\mu} \tag{6.4}$$

and the time it will take to reach some final temperature can be estimated from

$$t = \frac{KC_pm\Delta T}{P} \tag{6.5}$$

Trial-and-error evaluation can be minimized by using these equations to predict the conditions that will result from specific power and time exposures of a sample or samples. The analyst can also decide on an appropriate final temperature for a specific sample and use these equations to establish the correct power and time settings.

These thermodynamic relationships reliably predict temperature and time to within a few percent, for the mineral acids tested, despite small changes in their dielectric constants when the parameters are $\Delta T < 140\ °C$ or $t \leq 2$ min. In addition to acid decomposition, organic synthesis and other low-temperature applications can benefit greatly from this predicting capability (*21, 22*). At higher temperatures (150–250 °C) and longer exposure times (5–20 min), the actual conditions may deviate from their predicted values. A major source of error is heat loss through the walls of the container. Table 6-5 gives the deviation from predicted values for each of eight samples containing 5 mL of concentrated nitric acid and evaporated human urine standard reference material (SRM) 2670, (10 g each wet weight). Only the nitric acid couples significantly with the microwave power, so the mass of the sample is neglected in the calculations. Table 6-5 was calculated by using the heat capacity from Table 6-2, the power absorbed for eight samples [7.2 g each or 57.6 g (total) in 60-mL Teflon PFA vessels] of acid using the equations relating power consumption for nitric acid (equations 6.2 and 6.3), and the thermodynamic equation solved for time (equation 6.5). Various final temperatures (T_f) were used to calculate the theoretical time to reach a particular temperature at full (574 W) power; these times are compared with the actual elapsed time in Table 6-5.

The values agree within experimental error for up to 1.8 min, when a negative bias in the predicted value becomes apparent. This bias is caused by heat loss through the container walls. Such losses in Teflon PFA containers may reduce the effective power provided to the acid by as much as 50% in 6 min. At a given microwave power setting, the heat loss results in higher predicted temperatures than actually measured. Conversely, heat loss results in underprediction of time to reach a given temperature. The magnitude of both biases will increase with longer time or higher temperatures. For this type of Teflon PFA container, the reproducibility of this heat loss between lots under identical conditions is within several percent.

These deviations from predicted behavior vary with the heat loss of the vessel and thus become smaller with greater thermal insulation of the vessel. When uninsulated Teflon vessels are used, accurate use of these transformed equations is limited to the initial increase in temperature. Within these restrictions, the calculations provide a method of determining the amount of applied power necessary to reach a particular temperature or the approximate time this temperature is obtained at some applied power. However, no predictive information can be obtained about the amount of

Table 6-5. Predicted and Actual Time (s) for a Urine Sample to Reach a Target Temperature

Temperature (°C)	Linear	Quartic	Actual
110	64	69	54
130	80	86	72
150	95	102	81
160	102	110	102
170	110	118	113
180	117	126	144
183	120	129	150

power required to sustain the temperature. Such power requirements are dependent on the equilibrium that is established between the heat input and the heat loss at a given temperature.

Although the sample container does not absorb microwave energy directly, it does allow heat to flow from the sample. This condition establishes a thermal gradient inside the container that reduces the temperature of the volatilized acid below that of the liquid phase. Because the gas phase is cooled by heat loss from the vessel and absorbs little heat from the microwave field, condensation occurs at the top of the container above the liquid phase. Heat loss from the liquid phase is more than compensated for by its absorption of microwave energy. When a sample is placed in 5 mL of nitric acid in the 120-mL Teflon PFA vessel, more than 95% of the vessel volume is filled with the gas phase. Less than 5% of the volume is actually heated when the liquid phase absorbs microwave energy. Thus, the gas phase is not in thermal equilibrium with the liquid phase. This phenomenon prevents the use of partial pressure data accumulated under equilibrium conditions to predict the pressure inside the vessel at a given temperature. The actual pressure in a given container depends on the size and composition of the vessel, the type and quantity of acid(s) used for the dissolution, the temperature of the acid, and the temperature in the microwave cavity. The nonequilibrium conditions in the microwave vessel make direct measurement of pressure the only practical way of relating pressure to temperature.

Actual temperatures in the thick-walled Parr bomb probably closely follow the theoretical predictions over a much wider range of temperatures and for longer times, because the bomb is better insulated and has only a fraction of the heat loss of the Teflon PFA vessels. When the internal temperature of the Teflon PFA vessel was 132 °C, the temperature of the outer wall on the bottom was near 92 °C. However, when the inner temperature of the Parr microwave bomb was well in excess of 200 °C, the outer temperature was only 37 °C. The previous calculations should prove useful in designing specific conditions for this microwave digestion vessel. Real-time temperature measurements in the Parr vessel have not been obtained, because probes cannot be inserted without compromising the integrity of the protective outer casing.

These time and temperature predictions can be applied in the safe use of the closed-vessel microwave technique. Unsafe conditions can be prevented by calculating the predicted maximum temperature or minimum time to reach a given temperature.

Partial Power and Power Programming

Microwave systems can deliver variable power to the sample cavity by "time-chopping" the power to the magnetron at full power. A specific amount of reduced power per unit time may be delivered to the cavity to reach or maintain specific temperatures. Full-power settings are not always used to digest 250- to 500-mg biological and botanical samples in 5–10 mL of nitric acid. Instead, fractional power provides a more controlled method of heating. When using a partial power setting, one must know the number of watts absorbed at full power, assuming that the linearity of proportional power has been maintained. The equipment used to acquire the data presented here is linear within 1% over the entire power range. With all other parameters the same, a proportion of P equivalent to the partial power desired is used. For example, at 20% power (115 W) equation 6.5 becomes

$$t = \frac{KC_p m \Delta T}{0.20P} \tag{6.6}$$

Most commercial home appliances have a 10- to 15-s time-chopped duty cycle, that results in full power for several seconds and no power during the balance of the cycle. This alternate heating and cooling results in a sawtooth temperature curve. For multiple sample decompositions at long cycles, some containers may not be equally exposed to the microwave field. This is because of the inhomogeneity of the field. The magnetron of a laboratory unit should have a very short duty cycle (e.g. 1 s), so that when partial power is needed the magnetron is on for only a fraction of a second and off for the balance of the cycle. Rapid movement of the samples through the cavity (360°/20 s) produces more uniform power absorptions between multiple samples. The resultant heating curve is smooth without the wide variation in temperature that results from partial power settings. Multiple samples in such decompositions are likely to receive the same amount of microwave energy and, therefore, to experience similar heating. As seen in Figure 6.7,

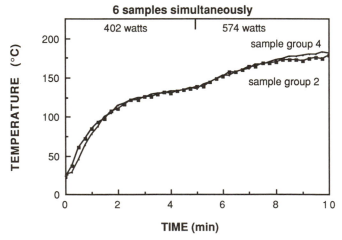

Figure 6.7. Comparison of the heating profiles of two different samples in two different sets of wheat flour (in nitric acid).

two sets of six identical samples have indistinguishable temperature profiles (less than ± 1.7% variation up to 175 °C) during a two-stage digestion program consisting of 5 min at 402 W followed by 5 min at 574 W. Temperature profiles can be precisely and accurately reproduced in a calibrated microwave system; thus, procedures can be transferred from one microwave unit to others.

Acids

Acid dissolution of a sample matrix is governed by many complex relationships that must be evaluated. The acid or combination of acids is chosen for its efficiency in decomposing the matrix. It is usually desirable for the acid to form a soluble salt with the metal ion of interest. For these reasons nitric, hydrochloric, and perchloric acids are widely used in sample preparation for analytical chemical analysis. Knowledge of the sample matrix and its major elements and compounds is essential for choosing the appropriate acid to ensure complete sample dissolution.

Combinations of acids are frequently used in analytical chemistry to dissolve a sample matrix. For example, hydrofluoric acid alone is inappropriate for the decomposition of botanical material. However, if a siliceous component is present, hydrofluoric acid is added to nitric acid to liberate trace elements that would otherwise remain trapped with the silica. These combinations must be chosen on the basis of the chemistry of the sample matrix (*23, 24*).

Another important consideration is interaction between the acid and the digestion container. Hydrofluoric acid should not be used in glass and quartz. When microwave energy is directly coupled with acid in closed microwave-transparent plastics, additional factors must be evaluated. Sulfuric acid, which has a high boiling point (339 °C), can melt most plastics, including Teflon PFA. Although most of the mineral acids traditionally used in decomposition are good microwave absorbers, other properties, such as the stability of the acid in the microwave field, its vapor pressure, and its interaction with other acids when used in combination, must be evaluated before attempting a closed-vessel digestion.

Nitric Acid

Decompositions with nitric acid are among the most common. Nitric acid is a strong oxidizing agent and is widely used for liberating trace elements from biological and botanical matrices as highly soluble nitrate salts. Nitric acid is one of the few acids that can be obtained in ultrahigh purity for very low-level analytical analyses. Because of its relatively low boiling point (120 °C), open-vessel nitric acid decompositions are time-consuming, often requiring temperatures >120 °C, or the addition of other strong oxidizing agents, such as peroxide or perchloric acid, to completely destroy a complex organic matrix.

Nitric acid behaves ideally under microwave energy excitation. In a closed container, nitric acid can reach 176 °C, at about 5 atm (Figure 6.8). This temperature is more than 50 °C above its boiling point. At this elevated temperature, substantial increases in oxidation potential are achieved, and these reactions proceed more rapidly (1). An example of rapidly heating the acid to extreme temperatures is shown in Figure 6.9, in which 3 mL of nitric acid is heated from 23 °C to over 175 °C in 2.5 min with 258 W of power.

Hydrochloric Acid

Concentrated hydrochloric acid is an excellent solvent for certain metal oxides and for metals that are oxidized more easily than hydrogen (25). Under high pressure and elevated temperature, many silicates and numerous other refractory oxides, sulfates, and fluorides are attacked by hydrochloric acid to produce generally soluble chloride salts (26).

Normally, concentrated hydrochloric acid is not used to digest organic materials, because it is not an oxidizing agent. Nevertheless, it is an effective solvent for basic compounds such as amines and alkaloids in aqueous solutions as well as some organometallic compounds. Hydrolysis of natural products with hydrochloric acid is a routine preliminary procedure for analysis of amino acids (26) and carbohydrates. Figure 6.10 shows the temperature and pressure curve of hydrochloric acid.

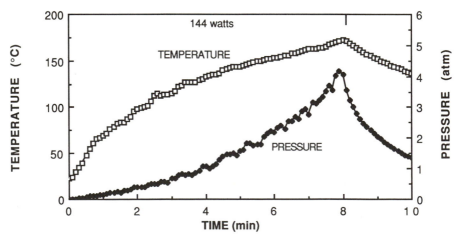

Figure 6.8. Temperature and pressure profile of 5 mL of nitric acid at 144 W power.

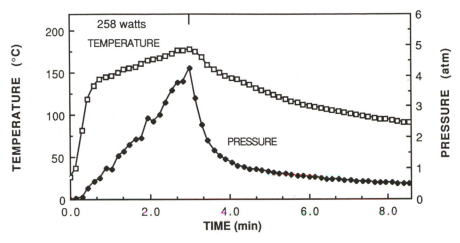

Figure 6.9. *Temperature and pressure profile of 3 mL of nitric acid at 258 W power.*

Hydrochloric acid may decompose to produce chlorine gas under high pressure or in the presence of strong oxidants (26). High-temperature digestions with this acid can be accomplished in Teflon PFA vessels.

Hydrofluoric Acid

Hydrofluoric acid is a useful reagent for dissolving silica-based materials. The silicates are converted to SiF_4, which can be volatilized and leave other

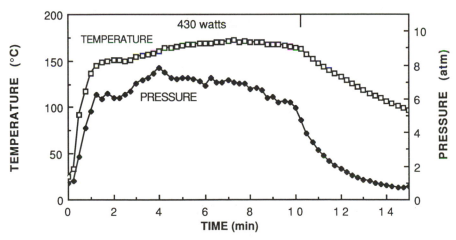

Figure 6.10. *Temperature and pressure profile of 7.3 g of hydrochloric acid at 430 W power.*

elements of interest. Hydrofluoric acid in small quantities is useful in combination with other acids to prevent silica from tying up trace elements in biological and botanical matrices. Because of its low boiling point (106 °C for 49% wt/wt) and high vapor pressure, hydrofluoric acid is easily volatilized and in closed containers has a partial pressure of nearly 8 atm at 180 °C. Frequently, quite stable metal–fluoride complexes are formed and can be kept in solution if the fluoride ion concentration is high enough. However, under these same conditions, rare-earth fluorides tend to be only sparingly soluble and may be lost from solution. The power consumption of hydrofluoric acid at 2450 MHz closely resembles that of nitric acid.

Phosphoric Acid

Hot phosphoric acid has been used successfully to digest iron-based alloys when hydrochloric acid would have volatilized specific trace constituents (27). Phosphoric acid will also dissolve a wide range of aluminum slags, iron ores, chrome, and alkali metals (26). The temperature and pressure profiles for phosphoric acid, shown in Figure 6.11, indicate that temperatures of 240 °C can be attained with just 3 atm. Because of its low vapor pressure, relatively high temperatures can be obtained without stressing the digestion vessel.

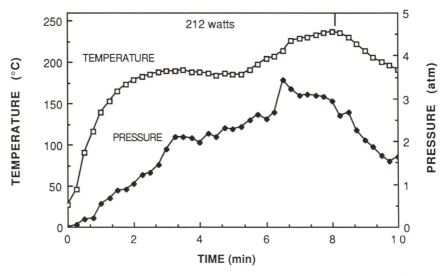

Figure 6.11. Temperature and pressure profile of 5 mL of phosphoric acid at 212 W power.

Tetrafluoroboric Acid

Tetrafluoroboric acid is used in some geologic decompositions (28) of inorganic matrices that require the attack of silicates and high temperatures. At 227 °C, the partial pressure of tetrafluoroboric acid in the closed vessel was only 5.7 atm. Temperatures much higher than can be used with hydrofluoric acid can be achieved without high pressures and the acid does not decompose.

Sulfuric Acid

Concentrated sulfuric acid is an effective solvent for a wide range of organic tissues, inorganic oxides, hydroxides, alloys, metals, and ores. Hot concentrated sulfuric acid can be used in Teflon containers, but its temperature must be monitored, because it is one of the few acids that can melt Teflon PFA before it boils. The use of glass and quartz vessels extends the useful temperature range of sulfuric acid.

Concentrated sulfuric acid completely destroys almost all organic compounds, and digestion times are reduced when the working temperatures are raised even a few degrees above its normal boiling point of 339 °C (98.3% wt/wt H_2SO_4). Using salts to raise the temperature of a sulfuric acid decomposition is especially advantageous in processes like the Kjeldahl analysis, in which a specific temperature is essential to the determination of nitrogen for accurate protein analysis (29). Figure 6.12 shows an open-vessel heating curve for sulfuric acid containing 0.5 g of protein and a catalyst.

Perchloric Acid

Hot concentrated perchloric acid is a strong oxidizing agent that attacks metals that are unresponsive to other acids. Perchloric acid also thoroughly decomposes organic materials. Because of its oxidizing capacity, the hot acid is frequently used to take elements to their highest oxidation state. Cold concentrated and hot dilute perchloric acid pose a reduced hazard; however, **hot concentrated perchloric acid is potentially explosive when in contact with organic materials and easily oxidized inorganics. Extreme safety precautions are required when using this concentrated acid at elevated temperatures.** Because of this potential hazard, expensive acid hoods, special scrubbers, and duct work are needed when working with perchloric acid. Discussions of the hazards of using perchloric acid can be found in the literature (30, 31).

An additional problem occurs when perchloric acid is heated in closed vessels in a microwave system. As seen in Figure 6.13, perchloric acid attains

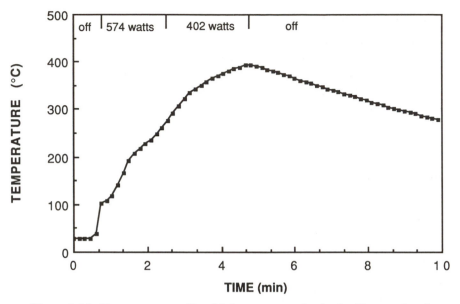

Figure 6.12. Temperature profile of 0.5 g protein in 5 mL of sulfuric acid with a catalyst in an open glass vessel.

temperatures >200 °C in <2 min at 212 W, with very little change in the initial pressure in the vessel. As the temperature approaches 240 °C, perchloric acid rapidly builds pressure with no increase in temperature until nearly 6 atm of pressure is generated. Figure 6.14 indicates that the temperature rise ceases as the pressure continues to build. Monitoring the cooling curve after power is terminated shows that the heating curve is not retraced, and that the vessel remains pressurized at room temperature. This finding suggests that perchloric acid does not simply vaporize when heated in a closed container, but undergoes an irreversible decomposition reaction that generates a gaseous end product when temperatures of 245 °C have been reached.

When NO_2 and CO_2 reaction products remain after nitric acid decomposition of organic tissue, liquid nitrogen cooling of the Teflon PFA reaction vessel normally relieves the pressure and allows safe cap removal (8). Figure 6.14 reveals that Cl_2, however, was not easily frozen out by this technique and 3.5 atm of toxic chlorine gas was left to vent. Because of this phenomenon, and because hot perchloric acid reacts explosively with organic matter, **perchloric acid should not be used in the microwave at this time.** The use of this acid under microwave heating conditions will have to be specifically studied. Accidents using conventional methods have occurred with mild heating conditions, so rapidly increasing the temperature of this acid clearly is a potential hazard.

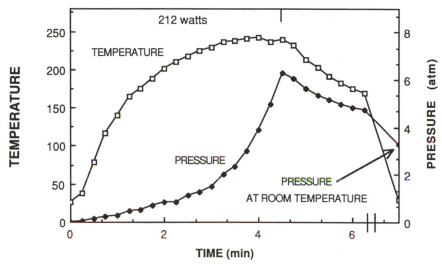

Figure 6.13. *Temperature and pressure profile of 5 mL of perchloric acid at 212 W power.*

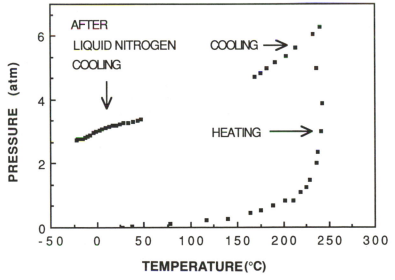

Figure 6.14. *Temperature vs. pressure profile of 5 mL of perchloric acid during heating and cooling.*

Mixed Acids

Temperature Calculation Capability and Heat Capacity

Acid combinations for microwave decomposition are practical for the same reasons they are used in wet-ashing procedures. Acid combinations are chosen for the ability of each acid to effectively decompose individual components of a particular matrix. In both open and closed vessels, combinations are frequently chosen because of the effectiveness of one of the acids as a digestion agent, and for the resultant aqueous solubility of the complexed elemental salts formed with a second acid during dissolution.

The calculation of temperature for an acid mixture requires knowledge of the heat capacity of that mixture. Most heat capacity data are for very dilute aqueous electrolyte solutions of a single mineral acid or salt. Heat capacities are not tabulated for acid mixtures. Therefore, mixed-acid systems have not been well characterized. Previous high-pressure acid dissolution techniques have included Carius tubes and steel-jacketed bombs, for which direct measurement of temperature and pressure was difficult. At present, the development of microwave power applications using mixed acids is restricted to actual measured results. Several mixed-acid systems have been evaluated in the microwave environment and are discussed in this section.

Pressure Considerations and the Temperature and Pressure Profile

The closed-vessel microwave dissolution technique has additional benefits when the acids are heated to relatively high temperatures in combination with each other. Nitric, hydrochloric, and hydrofluoric acids have relatively low boiling points and large accompanying partial pressures. Phosphoric, sulfuric, and fluoroboric acids have low partial pressures at comparable temperatures and have relatively high boiling points. By combining one acid from each group, these properties can be advantageously used to produce a mixture with a partial pressure somewhat less than that of the the lower boiling acid. Such acid mixtures are not ideal solutions and deviate from Raoult's law. The result of mixing acids is a useful lowering of the vapor pressures of the solutions.

Effective combinations that have been investigated in the microwave system include nitric and phosphoric acids for tissues, nitric and hydrofluoric acids for biological and botanical products, tetrafluoroboric acid with a nitric–hydrofluoric mixture for sludge, and aqua regia for mine tailings and geologic samples. Alpha-alumina has been successfully dissolved in 1:1 mixtures of phosphoric and sulfuric acids (26).

Nitric and Phosphoric Acids.   When certain acids are combined with nitric acid they significantly reduce the partial pressure of nitric acid. For instance, nitric–phosphoric acid mixtures (3:1, v/v) produce a combined vessel pressure of < 4 atm (4 × 10^5 Pa) near 180 °C, as seen in Figure 6.15. Nitric acid, however, at the same temperature is at slightly >5 atm (5 × 10^5 Pa) in the Teflon PFA vessel (nonequilibrium conditions).

Nitric and Hydrofluoric Acids.   Nitric acid in combination with hydrofluoric acid in a 3:5 or 1:1 ratio has temperature and pressure profiles, as shown in Figure 6.16 that are much like those of either nitric acid or hydrofluoric acid alone (compare Figure 6.8). The temperature vs. pressure curve of the acid mixture in Figure 6.17 increases gradually as the temperature is raised. The acids do not interact to produce decomposition products, as evidenced by the retracing of the cooling and heating curves.

Nitric and Hydrochloric Acids.   Aqua regia is a mixture of nitric and hydrochloric acids and is an effective oxidizing agent. Its effectiveness is due, in part, to the production of nitrosyl chloride (NOCl), one of the active agents in the mixture. On heating, nitrosyl chloride dissociates into toxic, highly corrosive chlorine gas. In addition to the chlorinating effect of the gases (24), aqua regia oxidizes many materials more efficiently than either hydrochloric or nitric acid alone. When confined to the container, these corrosive gases continue to attack the sample and produce an effective, closed-vessel aqua regia digestion. Freshly made aqua regia (1 HNO_3 : 3 HCl, v/v) at temperatures near 180 °C produces just over 7 atm (7 × 10^5 Pa) of pressure inside the vessel (Figure 6.18).

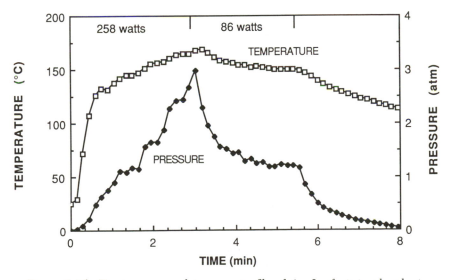

Figure 6.15. Temperature and pressure profile of 4 mL of nitric–phosphoric acid mixture (3:1).

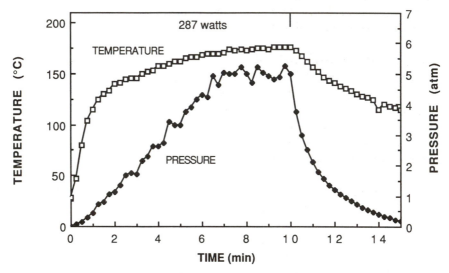

Figure 6.16. Temperature and pressure profile of two 8-mL samples of nitric–
hydrofluoric acid mixture (5:3) at 287 W power.

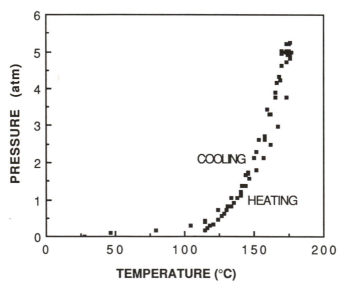

Figure 6.17. Temperature vs. pressure profile of 8 mL of nitric–hydrofluoric
acid mixture (5:3).

Examination of the temperature vs. pressure curves of a modified aqua regia (3 HNO_3 : 5 HCl, v/v) with an inorganic ore sample (Figure 6.19a) and of the same acid mixture without the sample (Figure 6.19b) shows that nearly identical conditions are reached. The coupling of the acid with the microwave energy is the major factor determining the heating profile and is essentially independent of the sample. The temperature vs. pressure profile of traditional aqua regia (Figure 6.20) shows a manageable heating curve, despite the tendency for gas formation by the mixture.

Sample Decomposition in Closed Vessels

Reactions by Specific Matrix Components in Biological and Botanical Samples

Rapid decomposition of biological and botanical samples that were digested in nitric acid was consistently observed. A one-gram sample of wheat flour (SRM 1567) and 10 mL of nitric acid were weighed into each of six 120-mL PFA vessels and were heated using 374 W of power (Figure 6.21). When the temperature reached 140 °C, 8 atm of CO_2 pressure from organic decomposition had accumulated in 1 min. The program was stopped after 3 min because of the pressure resulting from rapid decomposition. Figure 6.22 shows that this rapid decomposition is correlated with specific temperatures above the normal boiling point of concentrated nitric acid. Such behavior

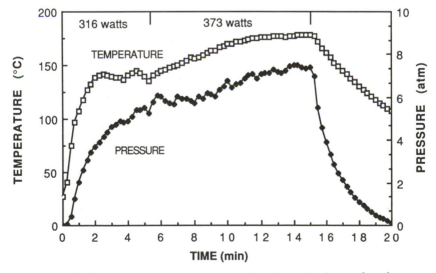

Figure 6.18. *Temperature and pressure profile of two 8 mL samples of aqua regia (1:3, nitric–hydrochloric, v/v).*

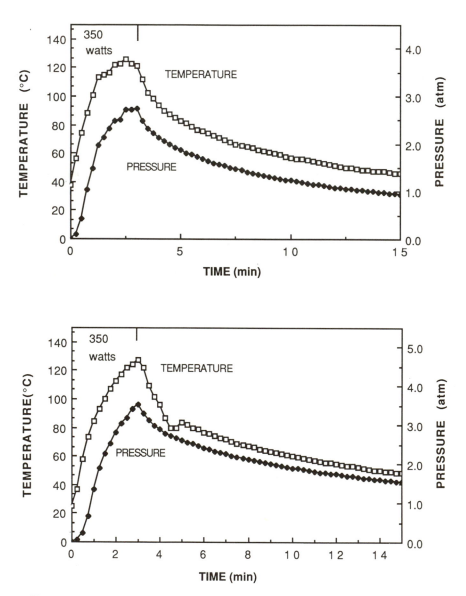

Figure 6.19. Top: Temperature and pressure profile of 345-mg zinc concentrate in 8 mL of aqua regia (3:5 nitric–hydrochloric, v/v) (18 samples). Bottom: Temperature and pressure profile of 8 mL of aqua regia (3:5 nitric–hydrochloric, v/v) (18 samples).

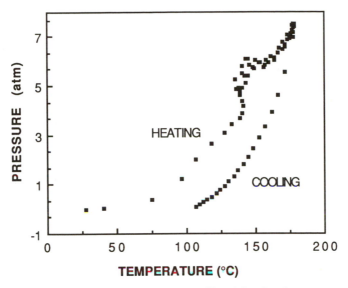

Figure 6.20. Temperature vs. pressure profile of 8 mL of aqua regia (1:3 nitric–hydrochloric, v/v).

was consistently identified with samples that have a high carbohydrate content.

Isolated materials representative of the carbohydrate in botanical tissue (two polysaccharides and one sugar monomer) were decomposed in the manner just described to determine whether the oxidation of the carbohydrate components of the botanical matrix was responsible for the sudden production of CO_2 in nitric acid at 140 °C. The polysaccharides were amylopectin–amylose (3:1) and soluble starch; glucose was the monomer. New, stronger 120-mL Teflon PFA vessels were used to withstand the partial pressure of nitric acid at 180 °C and the pressure produced by complete oxidation of 250 mg of organic material to CO_2. Figures 6.23–6.25 show the temperature vs. pressure decompositions of soluble starch in nitric acid, the mixed polymer, and the glucose monomer, respectively. A plot of pressure vs. temperature for this data emphasizes the dramatic trend in pressure production. All three graphs clearly show that the pressure increases dramatically at 140 °C and produces up to 10 atm of CO_2 with virtually no temperature increase. Finally, the temperature continues to rise with only a minor additional increase in pressure due to the partial pressure of nitric acid. These results demonstrate that carbohydrates decompose rapidly in nitric acid at 140 °C and that the carbohydrate component of a sample will decompose completely in 1–2 min under these conditions. The increased oxidizing power of nitric acid at elevated temperatures efficiently decomposes carbohydrates. Because nitric acid boils near 120 °C, this higher temperature cannot be reached in open vessels; thus, a closed-vessel microwave disso-

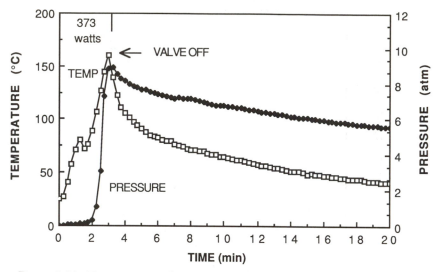

Figure 6.21. Temperature and pressure profile of 1 g of wheat flour SRM 1567 in 10 mL of nitric acid.

lution technique is much more efficient than traditional open-vessel dissolution.

Carbohydrates, proteins, and lipids are the three basic constituents of
biological and botanical tissue samples. These biological "building blocks"
were studied, and similar results were observed for each pure material. Bovine
serum albumin (SRM 926), (a 99% pure protein material), decomposes
rapidly in nitric acid at approximately 150 °C, whereas tristearin, a C-18
fatty acid ester, exhibited the same decomposition at approximately 160 °C.
When the closed-vessel nitric acid decomposition of a complete tissue sample, bovine liver (SRM 1577a), was compared with that of the biological
constituents, it was observed that the tissue decomposed readily at 160 °C.
Because the liver tissue contains carbohydrates, proteins, and lipids, thorough destruction requires a temperature that will decompose all of these
molecular species (*32, 33*).

Liver digestate was chromatographically separated to determine which
organic molecules remain after a 10-min nitric acid decomposition that
sustained temperatures of greater than 175 °C for 3–4 min. Only peaks for
o-, *m*-, and *p*-nitrobenzoic acid were found. The only organic bond not
decomposed under these conditions was the π bond of the benzene ring (*32,
33*). It is assumed that aromatic amino acids are the origin of these ring
structures (*32, 33*). These results indicate that the benzene ring will not be
decomposed by pure nitric acid under these conditions. Many inorganic
analytical determinations, such as inductively coupled plasma (ICP) or
atomic absorption spectrometry (AA), will not be seriously affected by trace
quantities of these nitroaromatics; other methods, such as polarography, will

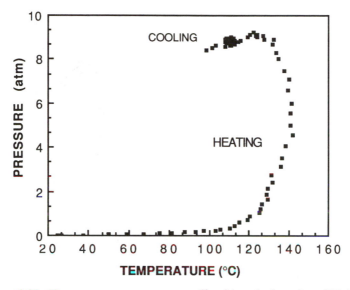

Figure 6.22. Temperature vs. pressure profile of 1 g of wheat flour SRM 1567 in 15 mL of nitric acid and water (2:1 v/v).

be severely affected. No significant traces of carbohydrates, proteins, or fatty acids were detected by high-pressure liquid chromatography (HPLC), a result suggesting that a 10-min digestion may be sufficiently complete for many techniques.

Such experiments demonstrate that knowledge of the constituents of a sample matrix is essential for determination of the most efficient temperatures for decomposition. Understanding the characteristics of a digestion and the interaction of components with specific reagents allows the analyst to more readily control the sample digestion process. Efficiently designed microwave digestions require reaching and maintaining the minimum temperatures that rapidly decompose the major organic components in the matrix. In the past, quality control in sample dissolution had been assessed by comparing the results of elemental analysis. Perhaps the specific characteristics of the digestion more properly form the basis for judgment prior to analysis.

Complete decomposition of organic matrices is only achieved with special reagents such as perchloric acid. Because this strong oxidizer should be avoided, higher pressure and temperature decompositions with nitric acid are an acceptable alternative. Microwave digestion of organic materials may require additional treatment with perchloric acid if complete decomposition is necessary before analysis.

The extent of nitric acid dissolutions of organic samples in steel-jacketed PTFE bombs at high temperatures and pressures has been studied (*1, 34, 35*). The results of these studies show that the decomposition of

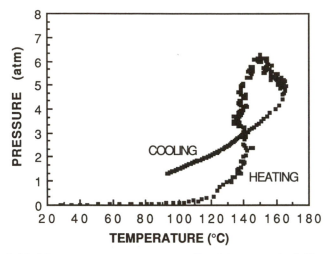

Figure 6.23. Temperature vs. pressure profile of 240 mg of soluble starch in 5 mL of nitric acid.

these samples was incomplete even after treatment at 180–200 °C for 3 h. When the residues from these dissolutions of biological tissue were analyzed, many of the same molecules were found in the PTFE bomb decompositions that had been found in the 10–15 min (180 °C) microwave decompositions. Similar ratios of *o-*, *m-*, and *p*-nitrobenzoic acids were found, as well as small amounts of dinitrobenzoic acid and several other aromatic compounds (35).

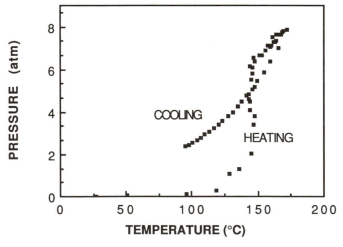

Figure 6.24. Temperature vs. pressure profile of 250 mg of amylopectin–amylose (3:1) polysaccharide in 5 mL of nitric acid.

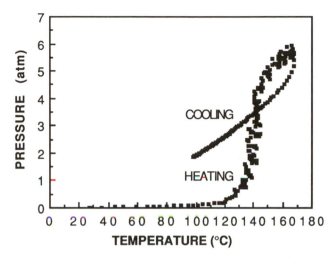

Figure 6.25. Temperature vs. pressure profile of 110 mg of glucose in 5 mL of nitric acid.

Nitric acid decompositions of organic samples are not complete even when high temperatures and pressures are applied for extended times. However, these high pressures and temperatures do appear to achieve reproducible decompositions. Under controlled conditions, nitric acid at high temperatures and pressures decomposes all but a few organic molecules. These molecules have been identified, in reproducible amounts, in solutions from dissolutions of both biological tissue and individual biological components. Because reproducible decompositions are possible with both microwave and steel-jacketed-bomb dissolutions, elemental analysis techniques that do not require complete decomposition of organic materials should be evaluated for their tolerance to these molecules. The closed-vessel, microwave, nitric acid, decomposition of organic tissue gives results comparable to traditional steel-jacketed PTFE bomb decompositions, but because of direct transfer of the microwave energy, the same temperature and pressure conditions as in the traditional digestions are achieved in a much shorter time. The usefulness of these high-temperature and high-pressure decompositions in conjunction with a variety of instrumental trace element analyses has been in the literature for several decades (1).

Despite the effectiveness of high-temperature and high-pressure decomposition, the specific chemistry of the acids' interaction with a particular matrix must be understood to optimize the dissolution. Microwave technology will not replace the need to understand the chemical reactions, many of which are accelerated by the elevated temperatures and pressures. It is necessary to know the acid(s) appropriate for the matrix, elemental analyte(s), and suitable temperature and pressure conditions for optimum dis-

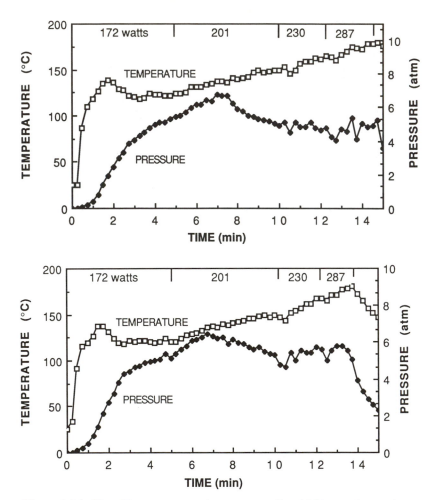

Figure 6.26. Top: Temperature and pressure profile of 250 mg of citrus leaves SRM 1572 in 5 mL of nitric acid. Bottom: Temperature and pressure profile of 300 mg of bovine liver SRM 1577a in 5 mL of nitric acid.

solution. Microwave technology should be implemented only after the optimum conditions have been determined.

Minerals, and possibly even alloy matrices, have similar ideal conditions under which they decompose most efficiently and reproducibly. Specific research into these decomposition conditions will reveal these parameters.

Benefits of Monitoring Acid Decomposition of Specific Sample Matrices

The acid decomposition of organic materials produces pressures in closed-vessel digestions that are higher than those that occur with highly oxidized

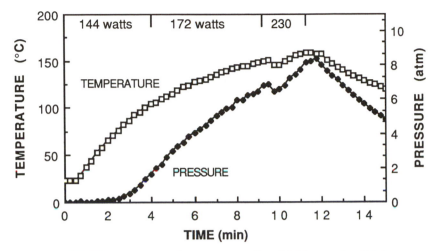

Figure 6.27. Temperature and pressure profile of 387 mg of molybdenum–chrome alloy in 8 mL of aqua regia (1:3 nitric–hydrochloric, v/v).

materials such as minerals, ores, ceramics, or some alloys. Monitoring the temperature and pressure is especially important to observe the added pressure from sample digestion products. Once the behavior of a material has been established, and the amount of the material that can be digested safely is determined, that material may routinely be decomposed with the confidence that repeatable pressures are being produced in the vessel.

Because the fundamental relationships and predictive (transformed) equations cannot be used to establish conditions in which the long-term maintenance of temperature or its slow gradual rise is desired, empirical data from the measurement of the temperature and pressure in the vessel must be used for methods development. Effective power consumption is reduced by the heat loss of the container and changes with each type of container used. Heat loss from a 60-mL Teflon PFA closed-digestion vessel is different from the loss in a 120-mL Teflon PFA vessel because the wall thickness of the latter is 40% greater. As new types of containers are used for closed- or open-vessel acid digestions, the heat loss for each type of container will be unique. By measuring the temperature rise with power over time for the acid or acid combination to be used in the particular type of vessel, the analyst can compensate for vessel variation by adjusting the input power to match each particular heat loss. A mathematical model is currently being investigated to establish this heat loss relationship for individual vessels. Combining this model with the other predictive relationships may enable the analyst to predict not only the amount of energy necessary to reach a particular temperature in a specified time, but also the energy requirements to maintain a narrow temperature range.

Multiple programming steps are frequently required to maintain, increase, or decrease acid temperature during digestion. Such programming can be accomplished as the conditions inside the vessel are observed. If the rate of temperature increase is not as fast as desired, or is not leveling off under some maximum target temperature, the program can be extended for 2 or 3 min by renewing the programmed cycle. Used in this way, the temperature and pressure measurements provide feedback information that allows the operator to exert a high degree of control over the reaction conditions. This technique is especially practical during research and methods development.

An example of the power programming technique is demonstrated in Figures 6.26a and 6.26b, in which the temperature is increased rapidly for 1 min and then increased slowly over the next 13–14 min. An apparent inconsistency in the decomposition of biological and botanical matrices is the small reduction in pressure observed after 8 min while the temperature continued to rise. This effect has been noted on many nitric acid decomposition profiles and appears to coincide with a sudden change in the solubility of NO_2, which is influenced by the total pressure in the vessel. At pressures >8 atm (8×10^5 Pa), the solution's color changes suddenly from yellow–brown to dark green. This increase in the solubility of NO_2 results in a decrease in pressure above the sample in the closed vessel. When the sample container is opened, the NO_2 slowly degasses, without effervescing, evolving the characteristic NO_2 fumes. The solution then returns to its more familiar yellow–brown color. The color change from green to yellow–brown is the result of nitric acid digestion products. It has been observed by chemists after decompositions that were carried out in Carius tubes and in steel-jacketed bombs. These color changes were probably not observed during decompositions before translucent Teflon PFA vessels were used for elevated pressures in microwave systems.

Another benefit of monitoring temperature and pressure is the ability to investigate the behavior of acid combinations or of previously untested solvents with unknown heat capacities or heating characteristics. The digestion conditions of an acid combination (e.g. aqua regia) can only be approximated before use. The exact conditions may not be predictable from theory, but a digestion can be modeled by measuring the temperature and pressure curves for this acid mixture. Monitoring the decomposition of a high molybdenum–chrome alloy in aqua regia shows that aqua regia behaves much like nitric or hydrochloric acid individually, but neither acid is a perfect analogue for the combination (Figure 6.27). These acid combinations are ideal for many sample matrices and can be monitored during procedure development to confirm the exact conditions under which the sample was decomposed before analysis. Once these conditions have been determined, the use of vessels capable of releasing gases at a specified pressure should allow reproducible decomposition of matrix components, without actually monitoring temperature and pressure. Because the major absorber of power, in most cases, is the acid and not the sample, many of the procedures that use the same quantity of the same acid should exhibit similar temperature profiles. Temperature and pressure measurement of decompositions with new reagents can provide empirical descriptions of expected conditions for which predictions cannot yet be made.

The power must be adjusted manually after evaluating the slopes of the temperature and pressure curves. In the future, automated equipment will allow the analyst to set the final temperature and an acceptable rate of change of temperature with time. A feedback system capable of continually adjusting the power to satisfy these preset conditions will be used to control the digestion. Temperature monitoring will certainly be necessary in such a system, as will the ubiquitous computer, to evaluate, calculate, project, and adjust the input power. As microwave dissolution matures, the sophistication of the equipment will improve and incorporate different methods of monitoring and power control. Until the automated equipment is more commonly available, the current, manual, successive power-per-time programming techniques are necessary to achieve these variable-staged conditions.

Safety Advantages of Monitoring Parameters

In addition to providing information for research or quality control, real-time monitoring of conditions is good laboratory practice. Because microwave energy is a directly coupled power source and is available at relatively high wattages, it is possible to create unsafe conditions in a very short period of time. Real-time monitoring provides the opportunity to observe the conditions in the vessels and to reduce or discontinue the power before it exceeds the vessel's specifications. The vessel temperature and pressure limits depend on the design of the container and the type and grade of material. Thus, these limits will be unique for each container model. Furthermore, if conditions are not monitored, there will not be enough warning to prevent unwanted or uncontrolled vessel venting.

Even if decompositions are monitored, using a pressure-relief valve can help to minimize the possibility of unsafe situations occurring. The form of pressure relief depends on the particular type of vessel and equipment configuration. If the release pressure of the valve is set to just below the safe operating limit of the vessel, the device will compromise the closed system only if the pressure becomes unsafe.

The absorption of power by a solution is influenced by the solution's concentration, ionic strength, and the molecular species present in the liquid phase. At any temperature, the pressure developed in the vessel depends on the partial pressure of the solvent(s) (under nonequilibrium conditions) and the gaseous digestion products formed during decomposition. Because these conditions are complex, it is difficult to exactly predict reaction mechanisms. Measurement of the pressure produced during these reactions aids in the investigation of microwave acid decomposition and other microwave and molecular interactions. Mechanisms of decomposition are reflected in the temperature and pressure curves that provide valuable information about the reactions. Through these real-time measurements during acid decomposition we have maintained a safe laboratory environment, have achieved some ability to control rapid reactions, and have begun to understand some of these complex interactions. On the basis of our experience we recommend the use of this methodology to other researchers who wish to study and perform microwave acid decompositions.

Acknowledgment

The authors gratefully acknowledge Susannah Schiller, of the Statistical Engineering Division of the National Bureau of Standards, for her effort and statistical expertise in modeling the microwave power absorption data and the production of these graphs.

Appendixes

These appendixes are the compilation of microwave power absorption measurements made on a variety of different mineral acids and water in a microwave system operating at 2450 MHz and a power output of 574 W. These data provide the basis for mathematical models that permit the prediction of absorbed power for specific masses of mineral acids in similar systems. The statistical uncertainties in these models are the 95% confidence limits. This value may be an underestimate of the uncertainty in the case of the first-order model in which the data may not be linear. Appendix A is the graphic presentation of the original scale data with a quartic fit surrounded by the confidence bands for prediction. The graphs have been informally arranged to group dilutions of the appropriate acid and to provide a family of curves as in the cases of HCl and HNO_3. Appendix B gives the full coefficients for the first-order and fourth-order models. When using the fourth-order model it is very important not to round off any of the coefficients provided for the quartic equation or any intermediate terms. Reducing the number of significant figures provided for the fourth-order coefficients may result in grossly inaccurate predictions. Appendix C is the graphic representation of the data on natural log scale with first and fourth-order models superimposed.

Appendix A: The Graphic Presentation of the Original Scale Data with a Quartic Fit Surrounded by the 95% Confidence Bands for Prediction

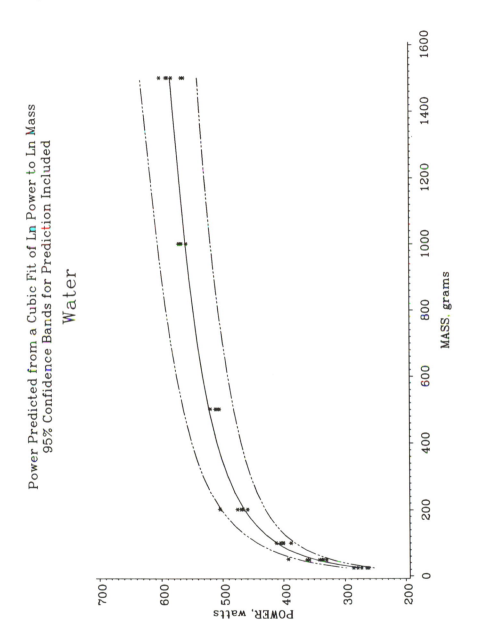

Water

Power Predicted from a Cubic Fit of Ln Power to Ln Mass
95% Confidence Bands for Prediction Included

Power Predicted from a Quartic Fit of Ln Power to Ln Mass
95% Confidence Bands for Prediction Included

Sulfuric Acid, 18M

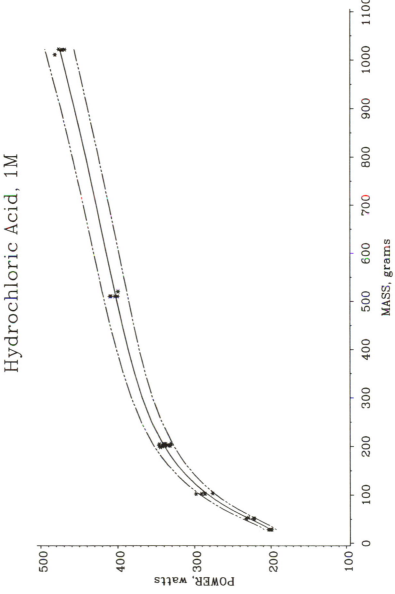

Power Predicted from a Quartic Fit of Ln Power to Ln Mass
95% Confidence Bands for Prediction Included
Hydrochloric Acid, 1M

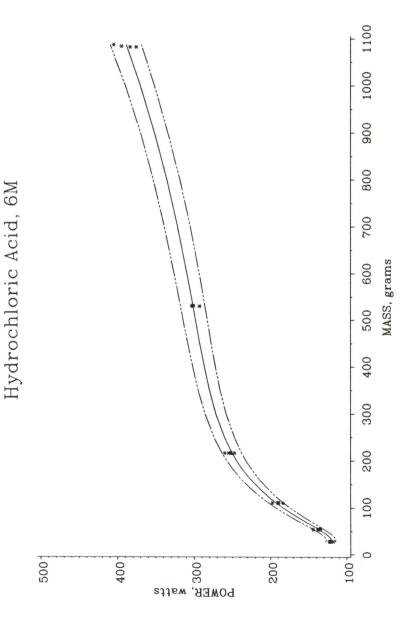

Power Predicted from a Quartic Fit of Ln Power to Ln Mass
95% Confidence Bands for Prediction Included

Hydrochloric Acid, 6M

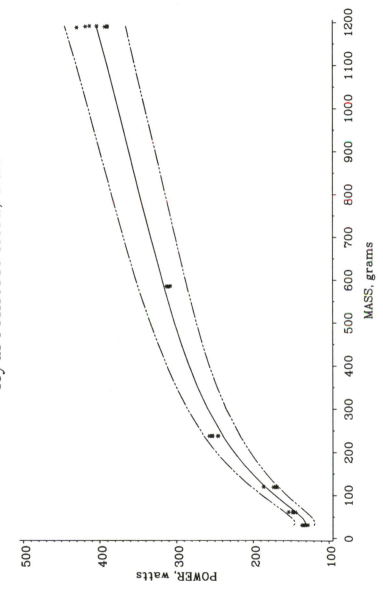

Power Predicted from a Quartic Fit of Ln Power to Ln Mass
95% Confidence Bands for Prediction Included

Hydrochloric Acid, 12M

Power Predicted from a Quartic Fit of Ln Power to Ln Mass
95% Confidence Bands for Prediction Included

Hydroflouric Acid, 29M

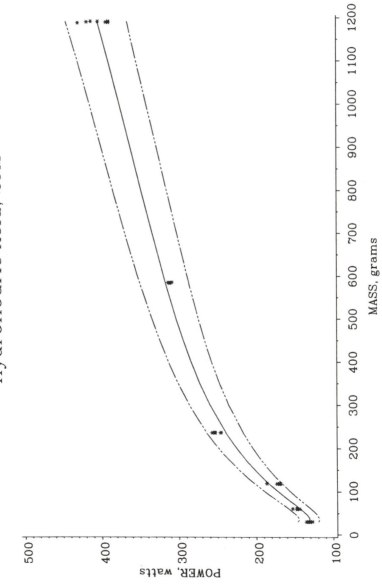

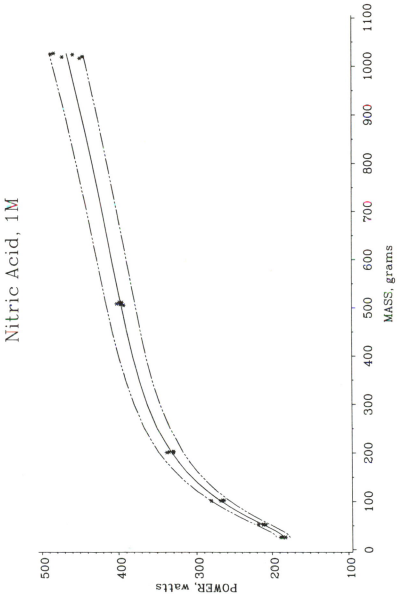

Power Predicted from a Quartic Fit of Ln Power to Ln Mass
95% Confidence Bands for Prediction Included

Nitric Acid, 1M

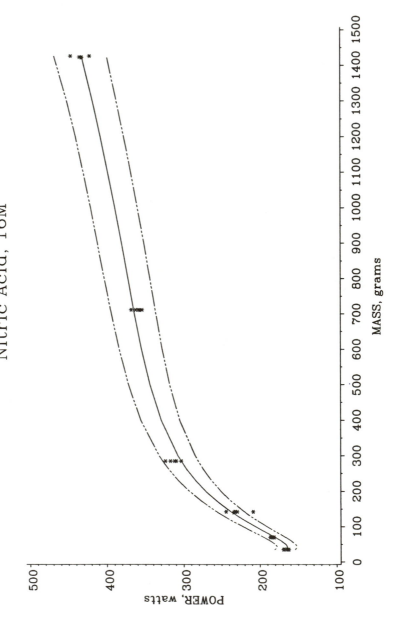

Power Predicted from a Quartic Fit of Ln Power to Ln Mass
95% Confidence Bands for Prediction Included

Nitric Acid, 16M

Appendix B: The Coefficients for the First-Order and Fourth-Order Models

Table I. First-Order Model

Acid	A'	B'
Water	5.1593378	0.17242256
H_2SO_4 (18 M)	4.36467137	0.25646561
HCl (1 M)	4.51149281	0.2421269
HCl (6 M)	3.70091195	0.32487487
HCl (12 M)	3.7240524	0.31907256
HF (29 M)	3.961217	0.30172725
HNO_3 (1 M)	4.39165309	0.25572899
HNO_3 (16 M)	4.11758785	0.27124904

NOTE: A' and B' are coefficients from equation 6.2.

Table II. Fourth-Order Model

Acid	A	B	C	D	E
Water	3.2003974	1.18320369	−0.160514	0.0079261	0
H$_2$SO$_4$ (18 M)	−8.29138	5.79513997	−0.666696	0	0.00270915
HCl (1 M)	14.0885888	−7.86067	2.49106325	−0.330417	0.01599077
HCl (6 M)	28.5892761	−20.2258	6.18708653	−0.806255	0.03845682
HCl (12 M)	19.5198223	−11.8319	3.42081805	−0.419341	0.01894251
HF (29 M)	36.4337146	−26.2649	7.91099431	−1.0178	0.0478557
HNO$_3$ (1 M)	16.8621705	−10.1777	3.16719797	−0.414305	0.01976774
HNO$_3$ − (16 M)	27.1480917	−17.3782	4.9371303	−0.598649	0.0261368

NOTE: A, B, C, D, and E are coefficients from equation 6.3.

Appendix C: Graphic Presentation of the Data on Natural Log Scale with First-Order and Fourth-Order Models Superimposed

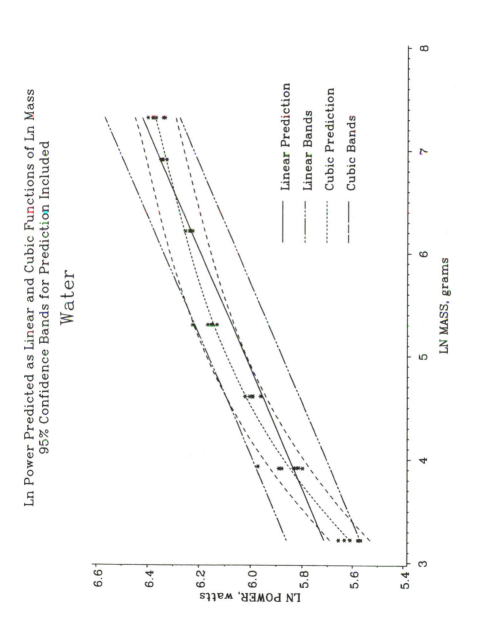

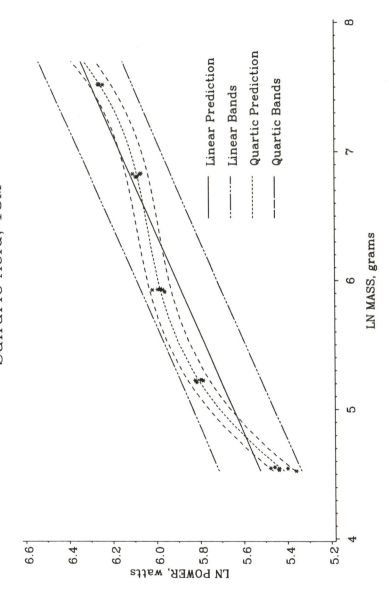

Ln Power Predicted as Linear and Quartic Functions of Ln Mass
95% Confidence Bands for Prediction Included

Sulfuric Acid, 18M

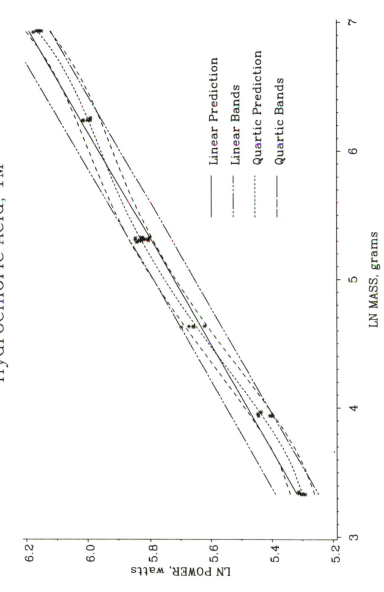

Ln Power Predicted as Linear and Quartic Functions of Ln Mass
95% Confidence Bands for Prediction Included
Hydrochloric Acid, 1M

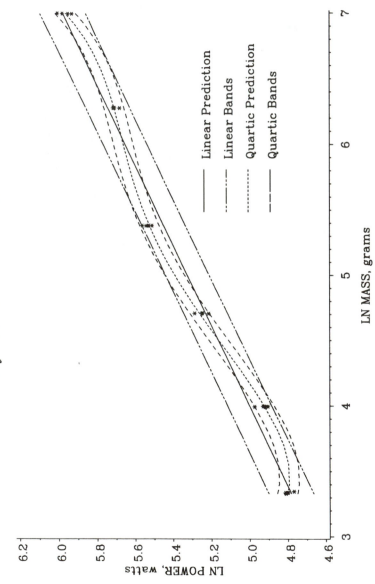

Ln Power Predicted as Linear and Quartic Functions of Ln Mass
95% Confidence Bands for Prediction Included

Hydrochloric Acid, 6M

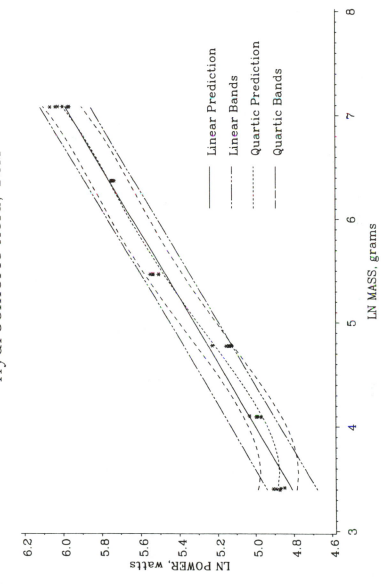

Ln Power Predicted as Linear and Quartic Functions of Ln Mass
95% Confidence Bands for Prediction Included
Hydrochloric Acid, 12M

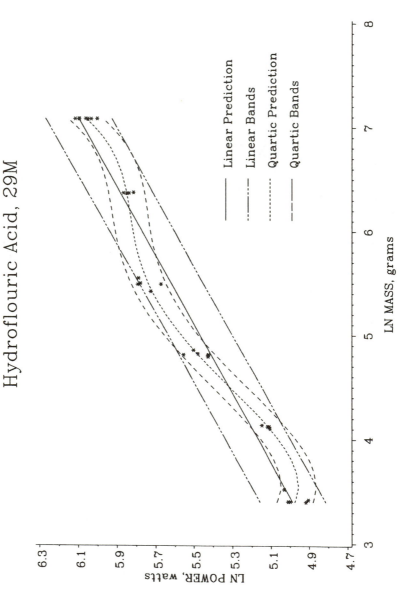

Ln Power Predicted as Linear and Quartic Functions of Ln Mass
95% Confidence Bands for Prediction Included

Hydroflouric Acid, 29M

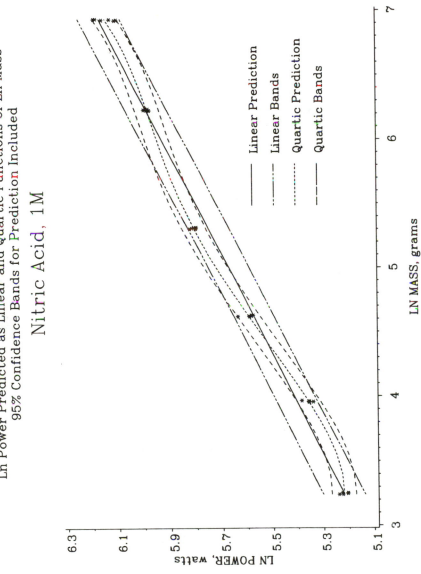

Ln Power Predicted as Linear and Quartic Functions of Ln Mass
95% Confidence Bands for Prediction Included

Nitric Acid, 1M

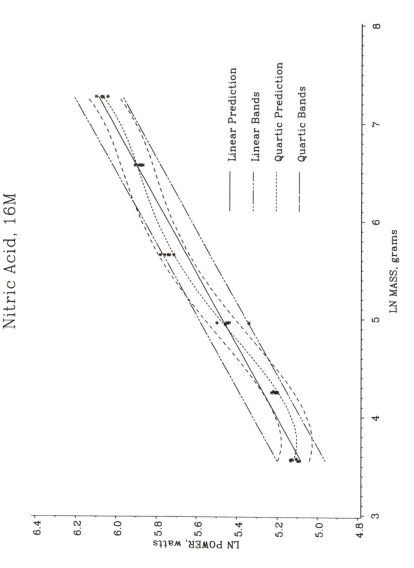

Ln Power Predicted as Linear and Quartic Functions of Ln Mass
95% Confidence Bands for Prediction Included

Nitric Acid, 16M

Literature Cited

1. Jackwerth, E.; Gomiscek, S. *Pure Appl. Chem.* **1984**, 56(4), 480–489.
2. Matthes, S. A.; Farrell, R. F.; Mackie, A. J. *Tech. Prog. Rep–U.S., Bur. Mines* **1983**, No. 120.
3. Kingston, H. M.; Jassie, L. B.; Fassett, J. D., Presented at the 190th National Meeting of American Chemical Society, Chicago, IL, September 1985, Paper ANYL 10.
4. Kingston, H. M.; Jassie, L. B. *Anal. Chem.* **1986**, 58, 2534–2541.
5. Sandberg, C.; Gerling, J. *Am. Soc. of Mech. Eng.* **1984**, 84-HT-50, 1–6.
6. Papoutis, D. *Photonics Spectra* **1984**, March, 5360.
7. Wickersheim, K.; Alves, R. B. *Industrial Research/Development* Dec., 1979.
8. Kingston, H. M.; Jassie, L. B. Presented at the 25th Eastern Analytical Symposium, October, 1986, Paper No. 76.
9. Sturcken, E., E. I. DuPont de Nemours, Savannah River Laboratory, Aiken, SC, personal communication, 1986.
10. Zakaria-Meehan, Z.; Neas, E., Presented at the 25th Eastern Analytical Symposium, October, 1986, Paper No. 75.
11. Minard, D. *Physiological and Behavioral Temperature Regulation*; Hardy, J. D.; Gage, A. P., Stolwijk, J. A. J., Eds.; Charles C. Thomas: Springfield, IL, 1970; Chapter 25.
12. Guy, A. W.; Lehmann, J. F.; Stonebridge, J. B. *Proc. IEEE* **1974**, 62(1), 55–75.
13. Lehmann, J. F.; Guy, A. W.; Stonebridge, J. B; Delateur, B. J. *IEEE Trans. Microwave Theory Tech.* **1978**, MTT-26(8), 556–563.
14. Johnson, C. C.; Guy, A. W. *Proc. IEEE* **1972**, 60(6), 692–718.
15. Copson, D. A. *Microwave Heating*; Avi: Westport, CT, 1975; p 443.
16. Gerling, E. E. *Microwave Energy Appl. Newsl.* **1978**, 11, 20–27.
17. Gerling J. E. *Gerling Laboratories Report*, **1980**, No. 80–013.
18. Watkins, K. W. *J. Chem. Ed.* **1983**, 60(12), 1043–44.
19. Parker, V. B. *Natl. Stand. Ref. Data Ser. (U. S. Natl. Bur. Stand.)* **1965**, NSRDS-NBS 2.
20. Parker, V. B. in *CRC Handbook of Chemistry and Physics*; 66th ed.; Weast, Robert C., Ed.; CRC: Cleveland, OH, 1985; p D–122.
21. Gedye, R.; Smith, F.; Westaway, K.; Ali, H.; Baldisera, L.; Laberge, L.; Rousell, J. *Tetrahedron Lett.* **1986**, 27(3), 279–282.
22. Gedye, R.; Smith, F.; Westaway, K. Presented at the 26th Annual Eastern Analytical Symposium, September, 1987, New York, NY, Paper No. 058.
23. Johnson, W. M.; Maxwell, J. A. *Rock and Mineral Analysis*; Wiley & Sons: New York, 1981; Chap. 4.
24. Dolezal, J.; Povondra, P.; Sulcek, Z. *Decomposition Techniques in Inorganic Analysis*; Elsevier: New York, 1968; Chap. 1.
25. Skoog, D. A.; West, M. W. *Fundamentals of Analytical Chemistry*; 2nd edition, Holt, Rinehart and Winston: 1969; p 756.
26. Bock, R. *A Handbook of Decomposition Methods in Analytical Chemistry*; translated and revised by Marr, I. L.; Wiley & Sons: New York, 1979; Chap. 4.

27. Kingston, H. M.; Paulsen, P. J.; Lambert, G. *Appl. Spectrosc.* **1984**, *38*(3), 385–389.
28. Matthes, S. Presented at the 25th Eastern Analytical Symposium, October 1986, New York, NY, Paper No. 72.
29. Neas, E.; Zakariah-Meehan, Z. *Introduction to Microwave Sample Preparation: Theory and Practice*; American Chemical Society; Washington, DC, 1988; Chap. 8.
30. Schilt, A. A. *Perchloric Acid and Perchlorates*; G. Frederic Smith Chemical: Columbus, OH, 1979.
31. Sax, N. I. *Dangerous Properties of Industrial Materials*; 5th ed. Van Nostrand Reinhold: New York, 1979.
32. Pratt, K. W.; Kingston, H. M.; MacCrehan, W. A.; Koch, W. F. Presented at the 193rd National Meeting of the American Chemical Society, April, 1987, Denver, CO, Paper No. ANYL 118.
33. Kingston, H. M.; Jassie, L. B. *J. Res. Natl. Bur. Stds.* **1988** 93(3), 269–274.
34. Stoeppler, M.; Muller, K. P.; Backhaus, F. *Fresenius' Z. Anal. Chem.* **1979**, 297, 107–112.
35. Wurfels, M.; Jackwerth, E.; Stoeppler, M. *Fresenius' Z. Anal. Chem.* **1988**, 330, 160–161.
36. *Lange's Handbook of Chemistry*; Dean, J. A., Ed.; McGraw–Hill: New York, 1979, 12th edit.
37. Kunzler, J. E.; Giauque, W. F. *J. Am. Chem. Soc.* **1952**, 74, 3472–3476.
38. Cox, J. D.; Wagman, D. D; Medvedev, V. A. *Codata–Key Values for Thermodynamics*; Hemisphere: Washington, DC, in press.

RECEIVED for review June 30, 1987. ACCEPTED revised manuscript June 3, 1988.

CHAPTER 7

Microwave Digestion of Biological Samples

Selenium Analysis by Electrothermal Atomic Absorption Spectrometry

K. Y. Patterson, C. Veillon, and H. M. Kingston

"Man's mind stretched to a new idea never goes back to its original dimensions".
Oliver Wendell Holmes

Closed vessel, microwave-heated digestions are used to rapidly destroy the organic matrix of biological samples with nitric acid at elevated temperatures and pressures. The system allows controlled, uniform application of microwave power and monitoring of temperature and pressure, and permits reproducible conditions while not exceeding the pressure or temperature limitations of the container. Samples are digested and analyzed for selenium by electrothermal atomic absorption spectrometry using matrix modification and Zeeman background correction. Analyte recoveries are established by using a radiotracer (^{75}Se) and accuracy is verified with standard reference materials of biological origin.

SELENIUM IS AN AN ESSENTIAL TRACE ELEMENT for rats, as has been known since 1957 (1). Since then, it has been shown to play a role in the activity of glutathione peroxidase in humans (2). The selenium content of soils, and therefore of foods grown on the soils, can vary widely. Finland, New Zealand, and portions of China are all known to have selenium-poor soil. In China, a selenium deficiency observed in the population of some rural areas has been linked to an increased incidence of Keshan disease (3). In contrast, in some areas in the United States, such as the San Joaquin Valley in California and portions of South Dakota, plants have accumulated enough selenium to be detrimental to the health of animals consuming them. This situation has made the development of accurate and reliable methods of analysis for selenium important.

Commonly used techniques for selenium analysis include fluorometry, hydride-generation atomic absorption spectrometry (HGAAS), neutron activation analysis (NAA), and electrothermal atomic absorption spectrometry (ETAAS). All of these are sensitive and can give accurate values in the nanogram-per-gram range. Except for NAA and analysis by ETAAS for some types of samples such as serum (4), the the sample must be digested before analysis by any of these techniques.

Fluorometry and HGAAS both require that the selenium-containing organic compounds be totally digested to allow for chelate or hydride formation. Several Se-containing compounds, such as selenomethionine, selenocysteine, and the trimethylselenonium ion, are difficult to digest fully. However, the use of high temperatures and strong oxidizing conditions results in the total breakdown of these compounds. In addition, if oxidizing conditions are not maintained throughout the digestion, samples may char; charring can result in reduced forms of selenium that are volatile.

Open-vessel, wet digestion procedures are usually used for Se in biological samples. One recommended method of preparing samples for fluorometric analysis uses nitric, perchloric, and sulfuric acids (5). Variations of this digestion, sometimes without H_2SO_4, are also used for preparation before HGAAS (6–9). The use of perchloric acid for any digestion requires special laboratory exhaust configurations and careful handling to prevent explosions.

Another method of wet digestion for HGAAS and gas chromatography–mass spectrometry (GC–MS), also requires complete digestion because the Se is chelated (10, 11). This method uses nitric and phosphoric acids along with hydrogen peroxide. Final temperatures of about 200–300 °C are used in this and in the previously described digestion procedures.

Analysis by ETAAS requires that the final sample solution contain the appropriate acids. Nitric acid is acceptable, but both sulfuric and phosphoric acid matrices are not because they generally give high background signals. The use of only nitric acid in an open digestion limits the temperature 120 °C (the boiling point of nitric acid), which is insufficient for the complete digestion of the selenium-containing organic compounds. Pressure digestion has been carried out by a number of researchers using nitric acid alone at 120 °C (12), 140 °C (13), and 160 °C (6). These authors report incomplete recovery of selenium on the basis of its determination by HGAAS.

A microwave energy-coupled pressure digestion procedure using only nitric acid in PFA Teflon [poly(tetrafluoroethylene)] vessels has been developed (14). Samples are rapidly heated to 180 °C; rapid heating gives a well-digested sample in a short time and in a matrix appropriate for ETAAS analysis. The loss of any volatile molecular forms of selenium is prevented by confining the vapor phase to the vessel during the digestion. Oxidizing conditions should be maintained far better during digestion in this vessel than in an open vessel. The nitric acid does not boil away and volatile reduced Se species are not formed or lost.

Development of Closed-Vessel Microwave Acid Digestion

Digestion Equipment

A microwave digestion system (CEM Corporation, MDS-81) was modified to allow both the temperature and pressure of a sample to be monitored during digestion. This equipment is described in detail in reference 14. Both 60- and 120-mL Teflon PFA sample vessels (Savillex Corporation) were used.

Organic Matrix and Microwave Acid Decomposition

When digesting organic materials in a closed vessel, considerable pressure is produced by gaseous byproducts, such as CO_2 and NO_2. It is therefore necessary to limit the amount of sample used and, particularly with the 60-mL vessels, to predigest the sample to remove the easily oxidizable material. This predigestion step reduces the total pressure.

   When the 60-mL Teflon vessels were used, it was necessary to reduce the pressure from the matrix during the microwave acid digestion by using a two-step digestion procedure. The samples were processed twice in the microwave system. The first step was predigestion, followed by cooling with liquid nitrogen to reduce the internal pressure before opening the vessel. The pressure and temperature during predigestion are shown in Figure 7.1.

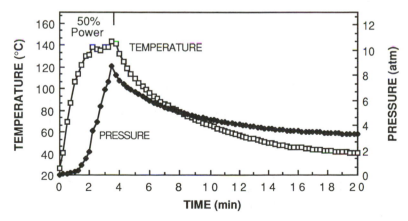

Figure 7.1. Pressure and temperature during microwave predigestion of a diet composite sample.

After the predigestion step, a second step at a higher temperature completed the decomposition as shown in Figure 7.2. The liquid nitrogen cooling was necessary to freeze excess CO_2; this freezing reduced the internal pressure of the vessel to atmospheric pressure or below (Figure 7.3).

The hot plate predigestion process was less reliable because it was done at lower temperatures. Moreover, depending on the specific matrix, the hot plate predigestion may not always be adequately energetic to destroy enough of the organic matrix to enable complete oxidation by a single microwave

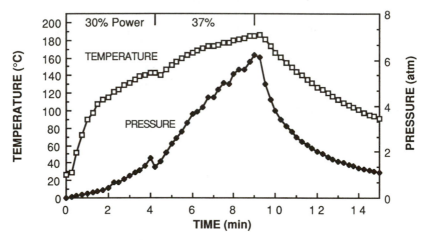

Figure 7.2. Pressure and temperature profile during digestion of a diet composite sample predigested in the microwave.

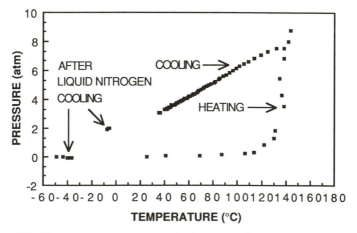

Figure 7.3. Temperature vs. pressure for heating and cooling during microwave predigestion of a diet composite sample.

digestion at high temperature. The success of this predigestion was dependent upon the experience of the analyst with regard to the particular matrix and the necessary experimental conditions. A successful predigestion can, however, enable direct use of the microwave system without overpressurizing the digestion vessels, as demonstrated by the pressures shown in Figures 7.4 and 7.5.

When the newer, specially designed 120-mL vessels are used for samples of 200 mg or smaller, a sample can usually be digested with a single heating

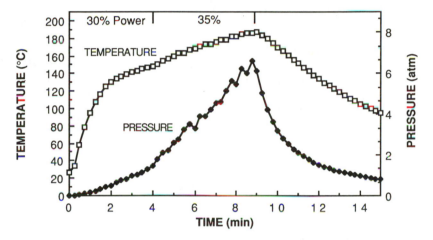

Figure 7.4. Pressure and temperature profile during digestion of a diet composite sample predigested on a hot plate.

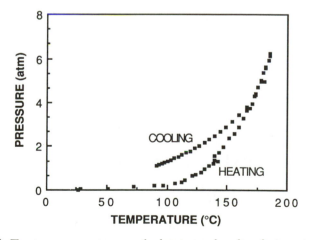

Figure 7.5. Temperature vs. pressure for heating and cooling during microwave digestion of a diet composite sample predigested on a hot plate.

in the microwave oven. The combination of the stronger design, thicker walls, and increased volume reduces the influence of the sample decomposition products. The 120-mL heavy-wall vessel has been used for up to 100 mg of a variety of biological and botanical samples, with single 180 °C nitric acid digestions. A predigestion step may be desirable if a 250-mg sample is used and nearly all the sample matrix is likely to form CO_2 or some other gaseous byproduct. The partial pressure of the acid relative to temperature remains the same.

Establishing Initial Microwave Conditions

The initial conditions necessary for nitric acid decomposition of organic tissue samples were derived from previous work (14). The organic material was placed in a closed Teflon PFA vessel and was kept at approximately 180 °C for several minutes. Knowing the total quantity of nitric acid in the samples to be analyzed we could use the fundamental microwave heating equation as described in reference 14 to calculate the time necessary to reach 180 °C. Usually, two-stage programs were used to compensate for the heat loss from the vessel.

Selenium Recovery Study Using Tracers

It is necessary to establish that no selenium is lost during any digestion procedure before the procedure can be confidently used for routine sample preparation. When oxidizing conditions are not maintained, selenium can be reduced and form volatile species. Another possible loss mechanism is adsorption by the walls of the digestion vessel. For example, one reduced form of selenium, SeH_2, has been reported (15) to be adsorbed by PFA Teflon. Although SeH_2 is not likely to form during digestion, other compounds that could be lost this way may be formed. It is relatively easy to follow radioactive isotopes through all steps in a proposed sample preparation procedure, because the gamma radiation can be measured without sample manipulation. For this reason, radioactive ^{75}Se was used as a tracer to monitor selenium during digestion. Two forms of selenium were studied: selenium as an inorganic salt mixed exogenously into an organic material, and endogenous selenium that had been incorporated into organic material, in a natural form. It is important to check for retention of both organic and inorganic forms because they may behave differently in a digestion.

Exogenous Tracer Study

Samples of a human diet composite (16) were exogenously spiked with 0.2 μCi (7.4 × 10³ Bq) of carrier-free radioactive [75]Se (as sodium selenite)

(New England Nuclear, Boston, MA). The diet composite consisted of foods that would represent a typical diet and is described in detail in Reference 16. The vessels were counted before preliminary open heating on a hot plate, after this predigestion, after the microwave digestion, and after samples were removed and the vessels rinsed with a small amount of water. The radioactive samples were counted in a small-animal whole-body counter (Nuclear Chicago Model 825) equipped with a multichannel analyzer (Canberra, Series 40) large enough to accommodate the 60-mL Teflon PFA vessels.

Six diet composite samples of 0.25 g each were placed in Teflon PFA vessels. Five of the samples were spiked with radioactive ^{75}Se and one was left unspiked so that it could be used to monitor temperature and pressure during the digestion. Nitric acid (5 mL) was added to each vessel and all were heated for 30 min on a hot plate at 60 °C. Because the older style, 60-mL Teflon vessels cannot withstand as much pressure as the new 120-mL redesigned vessels, two predigestion steps were added to prevent overpressurizing these vessels containing the radiotracer. The 120-mL vessels were not available for these experiments. After predigestion, the vessels were capped and the microwave oven was operated at 50% power (287 W). The pressure increased rapidly and was almost 9 atm (9 $\times$ 10^5 Pa) when the temperature reached 140 °C. The pressure and temperature recorded during the digestion are shown in Figure 7.1. The microwave was turned off at 4 min to prevent over-pressurizing the PFA vessels. Continued monitoring revealed that the internal pressure remained above 3 atm (3 $\times$ 10^5 Pa) when the vessels had cooled to room temperature. The vessels were cooled with liquid nitrogen, reducing their internal pressure below ambient, uncapped, and allowed to degas as they warmed. The temperature vs. pressure curve for the heating and cooling of these samples is presented in Figure 7.3. Three samples were recapped, and a second microwave digestion was used to reach the final temperature of 180 °C. These samples were allowed to cool to room temperature and were uncapped in a class-100 clean air hood. The temperature and pressure vs. time profile for the second digestion is shown in Figure 7.2. Microwave conditions were 4 min at 30% power and 5 min at 37% power. The difference between the pressure in Figure 7.1 and Figure 7.2 is caused by the digestion products of the organic matrix. Because most of the organic compounds were destroyed in the first microwave digestion, the pressure in the second was produced primarily by the partial pressure of nitric acid. The temperature vs. pressure curve for the second microwave decomposition (Figure 7.6) shows that after cooling, only a small residual pressure component from carbon dioxide production remains. This procedure, with both microwave digestions and liquid nitrogen cooling, can be completed in about 1 h. A hot plate acid decomposition procedure in Teflon beakers may take as long as 3 days to equivalently decompose the organic matrix.

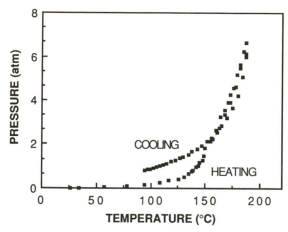

Figure 7.6. Temperature vs. pressure for heating and cooling during microwave digestion of a diet composite sample predigested in the microwave.

Endogenous Tracer Study

To check the retention through the digestion of organoselenium compounds, two rats were injected intraperitoneally with 3 μCi (1.1 × 10⁵ Bq) of carrier-free ⁷⁵Se, either as sodium selenite or selenomethionine (Amersham, Chicago, IL). After 24 h, the rats were sacrificed; serum, liver, and urine specimens collected during the previous 24 h were saved for analysis. The livers were freeze-dried before analysis. No loss of selenium was observed with this procedure (15). These endogenously labeled samples were counted in the same manner as the exogenously labeled samples.

The endogenously labeled radioactive rat serum, urine, and liver samples were prepared in a manner similar to that described for the mixed diet samples. Nonradioactive serum, urine, and liver from a control rat were also prepared so that temperature and pressure could be monitored. The 1-mL samples of serum and urine were dried before adding nitric acid. The liver samples weighed 0.25 g. The vessels containing the sample aliquots and acid were covered with watch glasses and heated on a hot plate for 48 h at 65 °C. The samples were brought back to their original weight with nitric acid, capped, and heated in the microwave unit to 180 °C.

Results of Tracer Studies

All the ⁷⁵Se exogenously added to the diet composite, was retained in both the samples that had been only predigested in the closed vessel in the microwave and in those samples that had been completely digested in the closed vessels in the microwave. No ⁷⁵Se was found on the lid or in the vessel after the sample had been removed and the vessel rinsed with a small amount of water. The uncertainty in this measurement is approximately 1–2%, a result suggesting that oxidizing conditions were maintained and that volatile species were not formed, or if formed, not lost. These results also indicate that the Se was not driven into the walls of the vessels by pressure.

The results of the digestions of the endogenously labeled rat serum, urine, and liver are given in Table 7-1. Selenium was not lost if the urine was evaporated only to dryness on a hot plate at 60 °C, but selenium was lost if the samples were heated beyond this point. There was a small but reproducible loss of selenium from the urine during the open-vessel predigestion step. There was, however, no additional loss during the closed-vessel microwave digestion, a result suggesting that it may be better to omit the predigestion step for urine. This omission is possible because the amount of organic material that produces CO_2 is relatively small in urine. No loss of selenium from serum or liver was observed for this digestion, even when the open-vessel predigestion was longer than necessary (48 h). The counting results on the small-animal whole-body counter are dependent on the geometry of the sample and the counting statistics. This probably accounts for the uncertainty in the retention values given in Table 7-1.

Verification of Complete Selenium Recovery

The applicability of the digestion method to the analysis of known samples was checked using samples of National Bureau of Standards (NBS) standard reference materials (SRMs) bovine liver (SRM 1577a) and wheat flour (SRM 1567), prepared and digested as previously described. The only difference was that the predigestion time was 1.5 h at 65 °C, and 120-mL, Teflon vessels were used. It was more convenient to use the hot plate for the predigestions than to heat the samples twice in the microwave oven.

A standard stock solution of 1000 mg of Se/L was prepared by dissolving elemental Se in a small amount of nitric acid (prepared by subboiling distillation) and diluting to volume with high-purity water.

Table 7-1. Retention of ^{75}Se from Endogenously Labeled Rat Serum, Urine, and Liver during Digestion

Material	Drying	Predigestion	Microwave Digestion	Vessel Walls
Serum	—	102% ± 2%[a]	101% ± 2%	<1%
Urine	99% ± 2%[b]	97% ± 2%	97% ± 2%	<1%
Liver	—	101% ± 2%	101% ± 2%	<1%

[a]Uncertainty is a combination of counting statistics and geometric changes.
[b]Samples cannot be heated after they are dry.

The digested SRM samples were diluted to 10 mL with high-purity water, and Se standards were made up concentrations of 0, 10, 20, 30, and 40 µg Se/L in 2% HNO_3. Analyses by ETAAS (Model 5000, Perkin-Elmer Corporation) used the Zeeman background correction technique. A matrix modifier solution was prepared to give 60 µg of Ni and 25 µg of $Mg(NO_3)_2$ per injection. With the autosampler, 20-µL aliquots of the samples were pipetted onto a platform in the graphite tube and an additional 5-µL aliquot of matrix modifier was added. The furnace program used is outlined in Table 7-2. Peak areas were recorded. The results were determined by comparison to standards and by the method of standard additions.

Both the liver and flour, with dry-weight concentrations of about 0.7 and 1.0 µg/g, respectively, gave peaks that were easy to distinguish from the background. The Ni in the matrix modifier was necessary to prevent loss of the Se before atomization, and the $Mg(NO_3)_2$ was added to the matrix modifier to improve the peak shape. A matrix effect from the inorganic components of the digest was evident because the standards and samples gave characteristic masses of 30 and 33 pg per 0.0044 A · s, respectively, which were comparable to literature values (17) but different from each other. The analysis was therefore made using the method of additions and the resulting values agreed with the certified values (Table 7-3).

The selenium determination by ETAAS for samples prepared by closed-vessel microwave acid digestion is satisfactory, provided the selenium con-

Table 7-2. Furnace Program used for Selenium Analysis of Standard Reference Materials

Program	Temperature (°C)	Ramp (s)	Hold (s)	Internal Flow (mL/min)
Dry	150	10	35	300
Char	900	10	20	300
Atomize	2100	0	4	stop
Cleanout	2700	1	4	300
Cool	20	1	14	300

Table 7-3. Analysis of Standard Reference Materials for Selenium

Sample	Certified Value µg/g	Experimental Value µg/g
Bovine liver	0.71 ± 0.07	0.70 ± 0.01
Wheat flour	1.1 ± 0.2	1.1 ± 0.1

NOTE: Values are ± uncertainity for certified values and ± standard deviation for experimental values.

centration of the samples is at least 250 μg/kg. This concentration results in digested samples that are sufficiently above the detection limit of 0.8 ng Se/mL to allow accurate analysis. The analytical blank for this procedure contains approximately 1 ng of Se or less.

Conclusions

The closed-vessel microwave digestion system and procedures described here offer a means of preparing biological samples for selenium determinations by electrothermal atomic absorption spectrometry. The organic matrix is destroyed without loss of analyte by volatilization or adsorption onto the container walls.

The closed, inert, microwave-transparent digestion vessels can be used for acid digestion at elevated pressures and temperatures. These vessels offer many advantages over open containers and facilitate the rapid destruction of the organic sample matrix. It is of utmost importance to control all variables (sample type, sample weight, acid volume, microwave power applied, etc.) during a digestion, and to monitor the pressure and temperature. This control allows reproducible digestion conditions and avoids exceeding the vessel's limitations. These biological, botanical, and composite materials represent a variety of organic samples that can be digested made suitable for selenium determinations by ETAAS with matrix modification and Zeeman background correction. Recoveries are verified by extrinsic tagging and by endogenously labeling samples with a radiotracer. The accuracy of the procedure has been verified by the use of SRMs.

Disclaimer

Certain commercial equipment, instruments, or materials are identified in this chapter to specify adequately the experimental procedure. Such identification does not imply recommendation or endorsement by the Department of Agriculture or National Bureau of Standards, nor does it imply that the materials or equipment identified are necessarily the best available for the purpose.

Acknowledgments

The authors thank Virginia Morris of the U.S. Dept. of Agriculture for her help with the animals and Lois Jassie of CEM Corporation for her assistance with the microwave digestions.

Literature Cited

1. Schwarz, K; Foltz, C. M. *J. Am. Chem. Soc.* **1957,** *79*, 3292–3293.
2. Awashthi, Y. C.; Beutler, E.; Srivastava, S. K. *J. Biol. Chem.* **1975**, *240*, 5144–5149.
3. Keshan Disease Research Group *Chin. Med. J.* **1979,** *92*, 477–482.
4. Lewis, S. A.; Hardison, N. W.; Veillon, C. *Anal. Chem.* **1986**, *58*, 1272–1273.
5. Analytical Methods Committee *Analyst* **1979**, *104*, 778–787.
6. Welz, B.; Melcher, M.; Neve, J. *Anal. Chim. Acta* **1984**, *165*, 131–140.
7. Salisbury, C. D.; Chan, W. *J. Assoc. Off. Anal. Chem.* **1985**, *68*, 218–219.
8. Mailer, R. J.; Pratley, J. E. *Analyst* **1983,** *108*, 1060–1066.
9. Piwonka, J.; Kaiser, G.; Tolg, G. *Fresenius, Z. Anal. Chem.* **1985**, *321*, 225–234.
10. Reamer, D. C.; Veillon, C. *Anal. Chem.* **1981,** *53*, 1192–1195.
11. Reamer, D. C.; Veillon, C. *Anal. Chem.* **1981,** *53*, 2166–2169.
12. Verlinden, M. *Talanta* **1982,** *29*, 875–882.
13. Welz, B.; Melcher, M. *Anal. Chem.* **1985**, *57*, 427–431.
14. Kingston, H. M.; Jassie, L. B. *Anal. Chem.* **1986,** *58*, 2534–2541.
15. Reamer, D. C.; Veillon, C.; Tokousbalides, P. T. *Anal. Chem.* **1981**, *53*, 245–248.
16. Miller-Ihli, N. J.; Wolf, W. W. *Anal. Chem.* **1986,** *58*, 3225–3230.
17. Manning, D. C.; Slavin, W. *Appl. Spectrosc.* **1983,** *37*, 1–11.

RECEIVED for review October 5, 1987. ACCEPTED revised manuscript January 19, 1988.

CHAPTER 8

Kjeldahl Nitrogen Determination Using a Microwave System

E. D. Neas and Z. Zakaria-Meehan

"It is the customary fate of new truths to begin as heresies and to end as superstitions".

Thomas Huxley

This chapter discusses recent developments in microwave dissolution of samples for the determination of total Kjeldahl nitrogen. The major difficulty with the Kjeldahl method is the first step, digestion. Conventional Kjeldahl digestion methods, instruments, and their limitations are presented. The microwave Kjeldahl method and instrumentation are also described. Digestion times and quantity of nitrogen obtained are compared for the two methods.

NITROGEN IS FOUND IN A WIDE RANGE OF BIOLOGICAL SUBSTANCES such as proteins, peptides, and certain vitamins. It also occurs in organic compounds such as amines, amides, and nitro compounds. The Kjeldahl method, named after its originator Johann Kjeldahl, is the procedure most frequently used to determine nitrogen content. This procedure, first described in 1883, has become one of the most widely used of all analytical procedures. There are three steps in the Kjeldahl analysis method:

1. The organic nitrogen is converted to an ammonium salt by digestion with concentrated sulfuric acid and additives (oxidants, catalysts, and salts in various combinations).

2. The ammonium salt is decomposed in basic solution to NH_3 that is distilled and recovered quantitatively.

3. The NH_3 is determined by titration with standard acid.

The following reactions illustrate sulfuric acid decomposition of various types of organic compounds (1).

1450–6/88/0167$06.00/0

Amides

$$C_{17}H_{35}CONH_2 + 53\,H_2SO_4 \rightarrow 18\,CO_2 + 52\,SO_2 + 69\,H_2O + NH_4HSO_4$$

Nitriles

$$C_2H_5CN + 8\,H_2SO_4 \rightarrow 3\,CO_2 + 7\,SO_2 + 8\,H_2O + NH_4HSO_4$$

Nitroparaffins

$$C_3H_7NO_2 + 7\,H_2SO_4 \rightarrow 3\,CO_2 + 6\,SO_2 + 8\,H_2O + NH_4HSO_4$$

Amino Acids

$$NH_2CH_2CH_2CH_2COOH + 10\,H_2SO_4 \rightarrow 4\,CO_2 + 9\,SO_2 + 12\,H_2O + NH_4HSO_4$$

The difficulty of the digestion arises from opposing objectives in the oxidation: the carbon must be oxidized to carbon dioxide, the highest valence form; at the same time, the nitrogen must be left as ammonium ion in the lowest valence state of the nitrogen (2). The Kjeldahl determination of nitrogen has been the object of numerous studies over the years, a situation reflecting not only the importance of the determination of nitrogen, but also the underlying difficulty associated with the method.

Despite the difficulties inherent in the Kjeldahl method, it is used to determine crude protein in food products, beverages, animal feeds, cereals, flour, grains, feedstuffs, yeast, and whey. Total Kjeldahl nitrogen, organic nitrogen, or percent nitrogen is determined in wastewater, coal, coke, fertilizers, and soil samples. The Association of Official Analytical Chemists has approved procedures for the Kjeldahl method that require a digestion time of about 2 h. This time requirement limits how quickly a nitrogen value can be obtained. In addition, the instruments used for the approved Kjeldahl digestion techniques consist of permanent installations, such as a fume hood, that occupy at least one wall of a laboratory, if not an entire room.

Microwave dissolution of samples for trace element analysis was first reported in 1975 (3) and has made a great contribution to that analytical procedure (4). This chapter discusses recent developments in microwave dissolution of samples for the determination of total Kjeldahl nitrogen.

Conventional Kjeldahl Digestions

With conventional Kjeldahl equipment, heat is applied externally to the digestion vessels by either a metal heating block (block digestor) (Figure 8.1), a flame, or electrical heating mantels. The block digestor is capable of heating up to 40 vessels at a time.

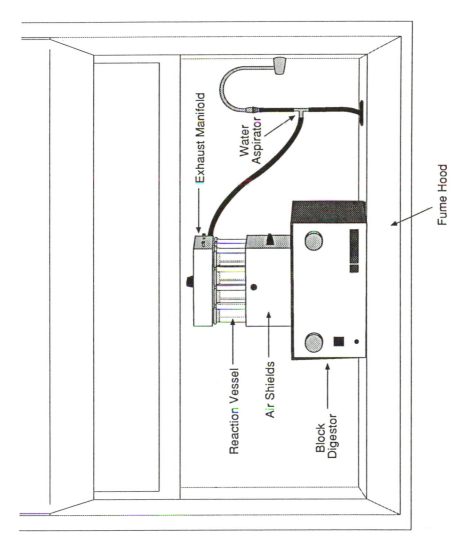

Figure 8.1. Block digestor apparatus.

The original method, the so-called standard Kjeldahl method, uses either a flame or an electrical heating mantel to heat the reaction vessels (Figure 8.2). From a user's standpoint, several limitations are associated with the use of conventional heating methods for Kjeldahl digestions. These limitations include aspiration of acid fumes and other noxious reaction gases, size limitations, external heat production, and reagent addition.

Methods

The traditional Kjeldahl digestion currently being used requires adding sample, catalyst, salt, and sulfuric acid to a reaction vessel and then heating the solution for 60–120 min. For the most part, the success of a Kjeldahl digestion depends on the temperature of the solution during the digestion process, because different samples require different reaction temperatures. The temperature at which sample dissolution occurs depends upon the type of nitrogen bonding and the number of those various bonds within a sample. In general, higher dissolution temperatures are required for samples that contain predominately heterocyclic and aromatic amines than for samples that contain mostly aliphatic amines.

Concentrated sulfuric acid boils at about 330 °C. To achieve temperatures above 330 °C, salt (typically potassium sulfate) is added to the acid to elevate its boiling point. The temperature reached during a digestion will depend on the acid-to-salt ratio (the acid index). The lower the acid index, the higher the final temperature of the digestion mixture. Figure 8.3 is a temperature versus time curve of a digestion mixture heated with the block digestor. The acid index of the digestion mixture was 1 (15 mL of H_2SO_4 : 15 g of K_2SO_4). It took 15 min to reach the boiling point of approximately 360 °C. After reaching the boiling point, there was a slow rise in temperature during the next 45 min for a final temperature of 400 °C. This rise in temperature is the result of evaporation of acid and the subsequent increase in the salt concentration (decreased acid index). The period of time during which acid is being volatilized is usually referred to as "after boil" and is an arbitrary time that is determined empirically according to sample type.

The length of the after-boil period determines how much acid is evaporated and, therefore, how high the final temperature will be. The after-boil time required for the complete conversion of nitrogen to ammonium sulfate will depend on the type of nitrogen bonding in the sample. However, there is a limit to the temperature at which dissolution is successful without loss of nitrogen. This limit varies according to sample type, catalyst used, and the acid index. A temperature much above 400 °C will lead to loss of nitrogen and low recoveries.

It is difficult to shorten the time required for a Kjeldahl digestion that uses conventional heating methods, although in certain instances it has been done (4). The approximate 15 min required to bring the solution to a boil

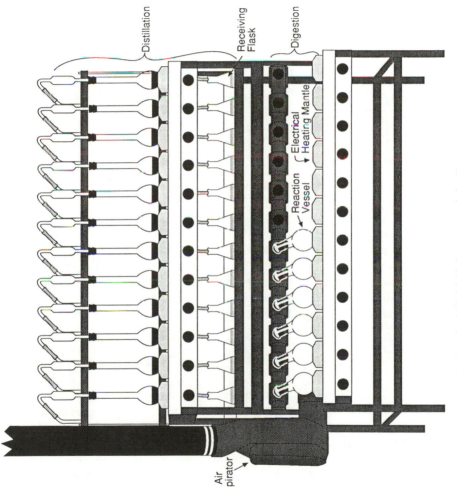

Figure 8.2. Standard Kjeldahl digestion and distillation apparatus.

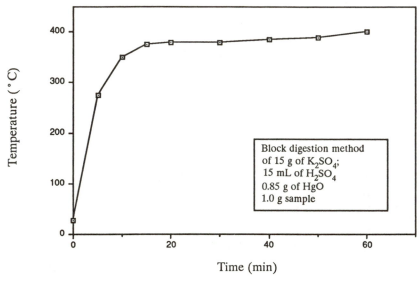

Figure 8.3. Temperature vs. time curve of a digestion mixture heated with a block digestor.

with a preset temperature of 400 °C is inherent to conventional heating methods and is discussed in Chapter 2.

The only way to decrease the heat-up time is to increase the externally applied temperature. However, this approach can lead to sample loss if the sample mixture boils over the top of the reaction vessel. "Boil-over" is the result of the initial reaction between the sulfuric acid and the carbohydrates of the sample,

$$C_{12}H_{22}O_{11} \rightarrow 12\ C + 11\ H_2O$$

which is called charring (2). Charring can be prevented by the early addition of oxidizing compounds such as hydrogen peroxide to the reaction vessel; however, strong oxidizing compounds pose potential safety hazards when being handled added by the user. A second problem that occurs with application of a higher external temperature is an increased rate of acid volatilization and, therefore, an increase in the rate at which the temperature rises. A high rate of temperature rise requires the analyst to remove the reaction flasks from the external heat source at a precise time. This is because reducing the heat source temperature (particularly for electrically heated metal blocks with large heat capacities) does not result in an immediate decrease in the applied temperature because some time is required for heat dissipation. Improper timing of the removal of the reaction vessels will lead to an excessive final temperature and subsequent loss of nitrogen or incomplete sample dissolution.

Instrument Limitations

Fume Aspiration. Two of the end products of a Kjeldahl digestion are SO_2 and CO_2. Most of the block digestors use water aspiration to remove these gases. As Figure 8.1 illustrates, a vacuum line runs from the water aspirator to an exhaust manifold on top of the reaction vessels. This method is adequate if there are no leaks between the exhaust manifold and the reaction vessels; however, this is not generally the case and most users have to place the block digestor in a chemical ventilation hood. The block digestor, therefore, occupies hood space needed for other laboratory work. Generally, the instruments used for the standard Kjeldahl method use air aspiration to remove the fumes from the reaction flasks (Figure 8.2). This aspiration method is adequate for fume ventilation but requires replacement of the aspiration system after a few years of use because of corrosion.

Size Limitations. The block digestion method was introduced in 1976 and represents the latest attempt to decrease the space restrictions associated with performing Kjeldahl digestions (7). The total bench space occupied by a complete block digestion system (block digestor and steam distillation system) is approximately 1–1.2 m (3–4 ft). The block digestor requires from 22 to 46 cm (9 to 18 in.) of counter top; this is the size required to hold from 6 to 40 reaction vessels. The block digestor is usually placed in a chemical ventilation hood and occupies valuable hood space.

The standard Kjeldahl digestion and distillation apparatus is shown in Figure 8.2. With all the equipment and accessories, it takes a whole room to use the standard Kjeldahl method for 12–18 samples at a time.

Generally, two types of steam distillation systems are used to separate NH_3 from a reaction mixture. One system only distills and collects the ammonia, so titration must be done separately by the analyst. A second type of system distills and automatically titrates for NH_3. Figure 8.4 shows an automatic steam distillation–titration system. This system occupies approximately 1 m (3 ft) of lab space.

External Heat Production. Both the block digestion method and the standard Kjeldahl method produce large amounts of excess heat. It is uncomfortable to be in the work area and expensive to adequately cool it.

Reagent Addition. As mentioned in the section on methods, it is tedious and hazardous to add oxidizing reagents to a conventional Kjeldahl digestion while the digestion is in progress. In order to add reagents, the analyst is required to handle hot reaction vessels. Also, when oxidants and water are added to the hot digestion mixture, a very violent reaction occurs and large amounts of gas are produced and may create a severe safety hazard.

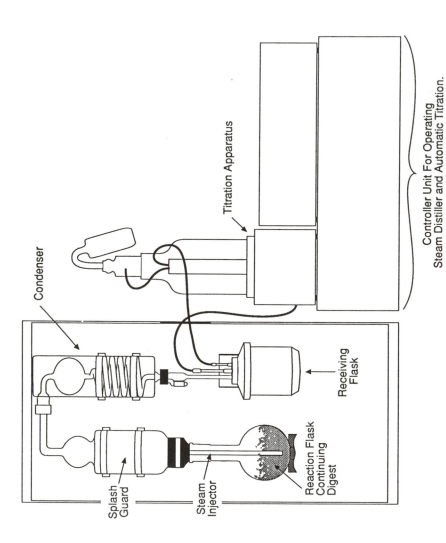

Figure 8.4. Automatic steam distiller and titrator.

Microwave Kjeldahl Method

Microwave energy is transmitted at the speed of light, passes through the walls of glass and quartz containers, and is rapidly absorbed throughout the acid–sample mixture. Sulfuric acid absorbs microwave energy very strongly so it heats quickly. During microwave heating, the digestion mixture's temperature rises to 360 °C in 2 min and to 400 °C after a total of 4 min. This rise is possible because energy is absorbed by the sample much faster than it is in the conventional methods. Because the rate of cooling is constant, the acid superheats slightly during microwave digestion. As soon as 400 °C is reached, microwave power is stopped. No hot mass is present to transfer heat to the sample, so heating stops instantly and temperature does not rise above 400 °C.

Instrumentation

Microwave energy has been used to develop a rapid, compact, and safe Kjeldahl method. The method allows a complete digestion and dilution for subsequent ammonia titration in approximately 6 min. The instrumentation that was developed is compact, requiring approximately 1 m² (3 ft²) of lab space, and, because of a self-contained fume scrubber, does not require a fume hood.

Figure 8.5 shows the microwave Kjeldahl instrument used for Kjeldahl digestions. The apparatus consists of three components: the microwave system, scrubber system, and pump and dispensing (P & D) system. Figure 8.6 shows the details of the microwave system. The reaction vessel is a 500-mL quartz flat-bottom boiling flask with a tapered ground-glass connector modified to seat a spherical joint. Quartz was used because its thermal expansion coefficient is almost zero; therefore, quartz enables the vessel to withstand the very rapid and extreme temperature changes that occur during water dilution (from 400 to 60 °C in ≈ 1.5 min).

A glass fiber insulator was developed to decrease heat loss through the walls of the reaction vessel during reaction. Glass fiber was used for insulation because it is relatively transparent to microwaves and can withstand the temperatures reached during the Kjeldahl reaction. The reaction vessel pedestal is spring loaded for insertion and removal of the reaction vessel.

The extension tube is made of quartz for the same reasons as the reaction vessel and has a spherical joint that fits inside the opening of the reaction vessel. The opening facilitates insertion and removal of the vessel. The extension tube leaves the microwave cavity through a microwave attenuator in the ceiling of the cavity. The attenuator prevents microwave leakage from the cavity.

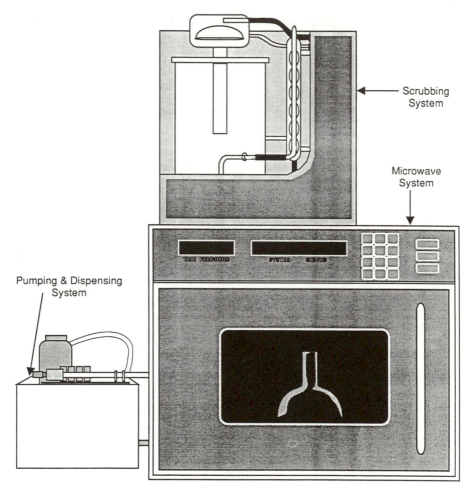

Figure 8.5. Microwave Kjeldahl apparatus.

Fumes from the reaction are drawn out of the microwave cavity by a venturi stream produced by aspiration of air through grooves in the spherical joint of the extension tube. The reaction fumes travel up through the extension tube, out through the fume outlet port on the distribution tube, to the cold water condenser, and then follow the pathway depicted in Figure 8.7.

Scrubbing System.   The self-contained scrubbing system is illustrated in Figure 8.8. The system uses a vacuum pump to draw air through grooves located on the spherical joint of the extension tube; thus a venturi flow is established for aspiration of the reaction fumes. The vacuum pump is a reed pump and has a flow rating of $0.019 \ m^3$ ($0.67 \ ft^3$). The heads of the vacuum

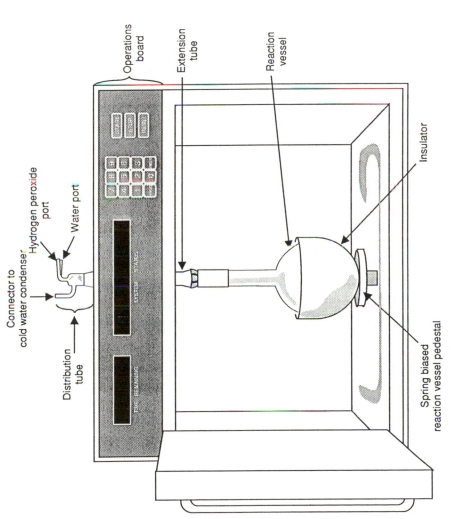

Figure 8.6. Microwave system.

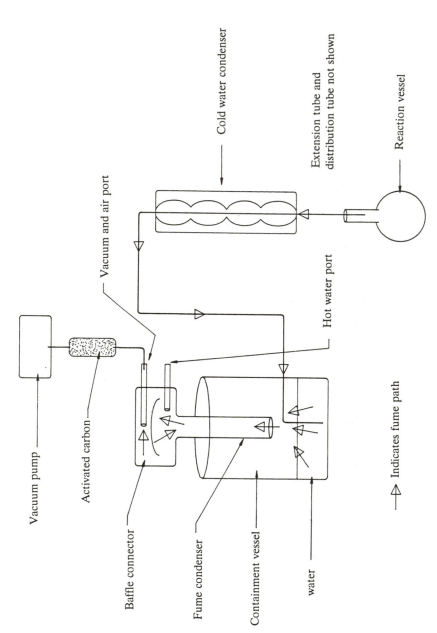

Figure 8.7. Fume ventilation path.

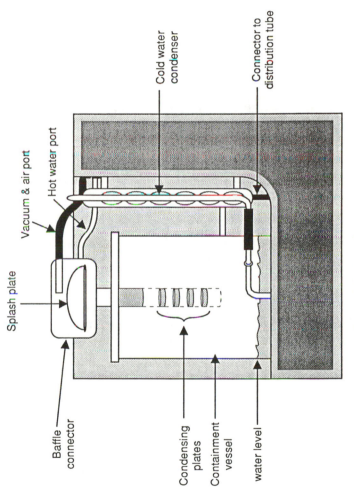

Figure 8.8. Scrubbing system.

pump are coated with and the reeds are made of fluoropolymer to prevent corrosion of the pump. To diminish corrosive fume contact with the vacuum pump, the scrubbing system contains a series of four glass filters that trap fumes between the water surface in the containment vessel and the filters. This trapping causes fumes to condense and dissolve in the water of the containment vessel. The condensate is automatically drained along with the containment vessel water at the end of each reaction. A small hot water heater in the scrubbing system provides hot water that is used to clean the filters after every reaction. It is necessary to clean the filters to remove fats that can evaporate during digestion and condense in the filters and clog them. The hot water is forced through the glass filters by a small pressure pump.

As an added safety precaution, the scrubbing system contains an activated carbon filter to sorb any sulfur dioxide that might escape through the filters. All the functions of the scrubbing system operate on command from a microprocessor located in the microwave system.

The Pump and Dispensing System.   Figure 8.9 illustrates the pump and dispensing (P & D) system. The P & D system pumps water and peroxymonosulfuric acid, H_2SO_5 (Caro's acid), to the reaction vessel through reagent ports on the distribution tube (Figure 8.6). The system also dispenses water to the condenser, hot water heater, and containment vessel on the scrubbing system. All water drainage from the scrubbing system is controlled through the P & D system. Line water pressure is regulated to 69 kPa (10 psi) by an in-line water flow regulator. A peristaltic pump is used to pump Caro's acid into the Kjeldahl reaction. A peristaltic pump is used to prevent highly corrosive Caro's acid from coming into contact with any pump parts. A polyethylene bottle is used to store the Caro's acid. As with the scrubbing system, the pump and dispensing system operate on command from a microprocessor located in the microwave system.

Methodology

The chemistry of microwave Kjeldahl digestion is basically the same as that of the standard Kjeldahl method. The method consists of adding a 1-g sample, 15 mL of sulfuric acid, the catalyst of choice, and 15 g of potassium sulfate to the reaction vessel. The difference in the chemistry is the addition of Caro's acid during the digestion process. As digestion proceeds, Caro's acid is automatically pumped into the reaction vessel three times according to a preset program designed by the user. Caro's acid has an oxidation potential of 1.81 V and rapidly oxidizes the hydrocarbons in the sample

$$C_{12}H_{22}O_{11} + H_2SO_5 \rightarrow 12\ CO_2 + 11\ H_2O + 24\ H_2SO_4$$

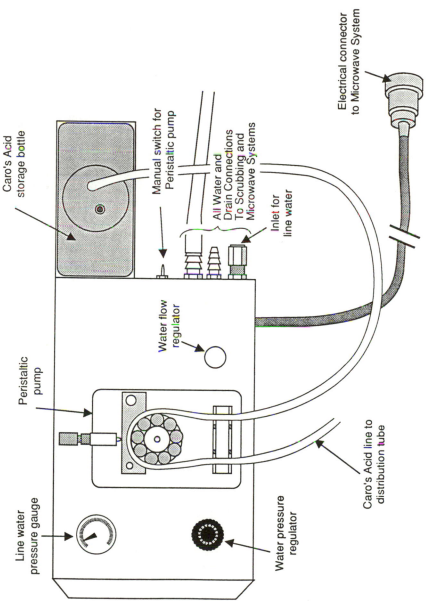

Figure 8.9. Pump and dispensing system.

Rapid oxidation of hydrocarbons by Caro's acid prevents charring of the sample that would cause boil-over and the subsequent loss of nitrogen. Also, the use of Caro's acid dramatically decreases sulfur dioxide production. Sulfur dioxide would be produced in large quantities if sulfuric acid were the only oxidizing agent in the digestion. Because Caro's acid is automatically added to the reaction by the microprocessor-controlled peristaltic pump, there is no risk to the user during the addition.

Figure 8.10 illustrates a typical time–temperature curve for the microwave Kjeldahl digestion of soy protein using the chemistry described in the previous paragraph. It takes approximately 90 s to reach the boiling point of the solution (370 °C). Another 150–210 s is required to reach the final temperature of approximately 400 °C. As in the conventional Kjeldahl methods, the rise in temperature is the result of acid loss through volatilization and the subsequent increase in salt concentration. Acid loss is more rapid in the microwave method because of the venturi effect of the scrubbing system that decreases the condensate return to the reaction flask. The obvious difference between convective heating and microwave heating is the time required to reach the optimum temperature for a complete digestion.

The sample matrix has an insignificant effect on the temperature rise during a microwave Kjeldahl digestion of solid samples (Figure 8.11). This

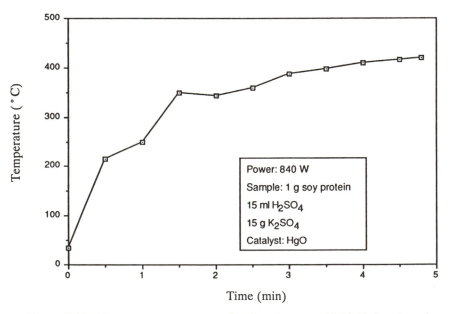

Figure 8.10. Time–temperature curve for the microwave Kjeldahl digestion of soy protein.

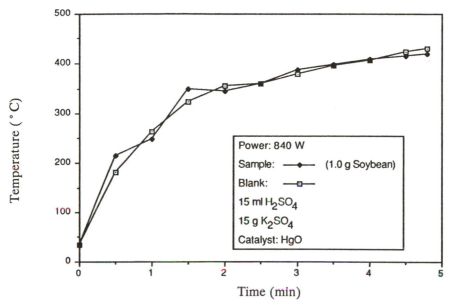

Figure 8.11. Time–temperature curves for a microwave Kjeldahl digestion with a soy bean sample and without a sample (blank).

insignificant effect is to be expected, because the sample mass is only 6% of the total mass of the acid–sample mixture and the sample contains a very small percentage of the absorbing dielectrics. Kingston et al. (6) demonstrated that nitric acid is the primary microwave absorber in acid dissolutions of samples for trace metal analysis.

Figure 8.12 compares the digestion times for the microwave Kjeldahl digestion method and the two most commonly used conventional methods. The microwave method is much faster. After microwave digestion, the ammonia concentration was determined with an automatic steam distiller–titrator system. A percent protein or total Kjeldahl nitrogen number (TKN) for a sample can be determined in less than 15 min with an automatic steam distillation–titration system used in conjunction with the microwave Kjeldahl digestion.

Table 8-1 compares the percent nitrogen obtained with the microwave Kjeldahl digestion method and the block digestion method. The microwave Kjeldahl method is comparable to the block digestion Kjeldahl method with all sample types tested. Although only a few different samples are compared, the microwave Kjeldahl method should be comparable to other methods with any sample type. This comparability is to be expected because the chemistry of the microwave Kjeldahl method is essentially the same as in other accepted Kjeldahl methods.

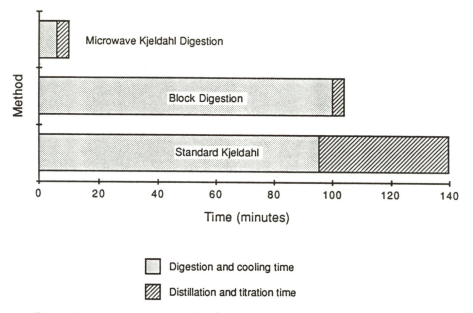

Figure 8.12. Comparison of the digestion times for the microwave Kjeldahl digestion, block digestion, and standard Kjeldahl digestion methods.

Summary

With microwave heating for sample preparation, the optimum digestion temperature can be reached very rapidly, and held long enough to complete the digestion, and the solution can be cooled immediately to avoid loss of nitrogen. A microwave Kjeldahl digestion can be as much as 20 times faster than a conventional Kjeldahl digestion. The precision and accuracy of the microwave Kjeldahl method is comparable to other accepted Kjeldahl methods. The microwave Kjeldahl system requires little laboratory space and is

Table 8-1. Kjeldahl Nitrogen Determination by the Digestion Block Method (DBM) and the Microwave Kjeldahl Method (MKM)

Sample	DBM	MKM	Difference
Nicotinic acid *p*-toluenesulfonate	4.74	4.74	0.0
Ground beef	2.96	2.97	+0.01
Franks	2.43	2.45	+0.02
Ground ham	2.60	2.64	+0.04
Sausage	2.21	2.21	0.0
Pastrami	3.06	3.06	0.0
Fish meal	10.35	10.37	+0.02

NOTE: All results are percent nitrogen.

simple to operate. Because of the built-in fume scrubber, the microwave Kjeldahl instrument can be used anywhere in a laboratory without the need of a fume hood.

The microwave Kjeldahl method has some distinct advantages over the currently used conventional methods, not the least of which is speed. One area of future research involves the use of a temperature feedback mechanism to control the digestion temperature. By using controlled temperature, the precision of the digestion should be improved. Another research area is the combination of microwave heating and microprocessor-controlled addition of Caro's acid to complete a Kjeldahl digestion without the use of catalyst. Digestion without a catalyst would be advantageous because many of the catalysts used in conventional Kjeldahl methods constitute a hazardous waste.

Literature Cited

1. Bradstreet, R. B. *The Kjeldahl Methods for Organic Nitrogen*; Academic: New York, NY, 1965; p 17.
2. Diehl, H. *Quantitative Analysis: Elementary Principles and Practice*; Oakland Street Science: Ames, Iowa, 1970; p 174.
3. Abu-Samra, A.; Morris, J. S.; Koirtyohann, S. R. *Anal. Chem.* **1975** *47*, 1475–1477.
4. Kingston, H. M.; Jassie, L. B. *Anal. Chem.* **1986**, *58* 2534–2541.
5. Suhre, F. B.; Carrao, P. A.; Glover, A.; Malanoski, A. J. *J. Assoc. Off. Anal. Chem.* **1982**, *65*, 1339–1345.
6. Nebergall, W. H.; Schmidt, F. C.; Holtzclaw, H. F. *College Chemistry with Qualitative Analysis*; 1963; p 364.
7. Noel, R. J.; Hambleton, L. G. *J. Assoc. Off. Anal. Chem.* **1976**, *59*, 134–140.

RECEIVED for review May 17, 1988. ACCEPTED revised manuscript July 2, 1988.

CHAPTER 9

Remote Operation of Microwave Systems

Solids Content Analysis and Chemical Dissolution in Highly Radioactive Environments

E. F. Sturcken, T. S. Floyd, and D. P. Manchester

"The frontiers are not east, west, north or south, but wherever a man fronts a fact".
Henry David Thoreau

Microwave systems allow the quick and easy determination of the solids content of highly radioactive samples in shielded cells. Chemical dissolution by microwave systems is also faster, by an order of magnitude, than previous methods that use remote manipulators. The modifications of a moisture–solids analyzer microwave system and a drying–digestion microwave system for remote operation in high-radiation environments are described. The use of these systems to determine solids content and to dissolve nuclear waste slurries is also described. These microwave systems have operated satisfactorily for 2 years in a γ-radiation field of 1000 R.

WITH ANALYTICAL INSTRUMENTS, it now takes longer to prepare the sample than to run the analysis. Hence, microwave ovens are being used for drying and chemical dissolution to increase the speed of sample preparation. The microwave instruments have become more sophisticated with the addition of microcomputers, electronic balances, high-pressure vessels to increase chemical reaction rates, and inserts for ashing or melting samples.

This chapter describes the modification of microwave instruments for remote operation in highly radioactive environments. High levels of radioactivity deteriorate the physical, chemical, and electrical properties of some of the materials used to construct the instruments.

1450–6/88/0187$06.00/0

Procedures are described for using the modified instruments in high-level radiation cells (HLC). These instruments are used for solids analyses and dissolution of nuclear waste at the Savannah River Laboratory; they will serve as prototypes of future instruments that will be used for process-control analyses in the Defense Waste Processing Facility that is being constructed for the vitrification of high-level radioactive waste (1) at the Savannah River Plant.

The microwave systems that were modified are manufactured by the CEM Corporation. One of the units, the AVC-80 (Figure 9.1), is designed for analysis of solids or moisture in materials. The other unit, the MDS-81 (Figure 9.2), is designed for chemical dissolution or ashing of materials. Addition of remote-operation capabilities to electronic components and mechanical modifications were engineered by Floyd Associates.

Design Strategy

The primary reason for remote operation of instruments is to protect personnel from harmful radiation; hence, the microwave ovens are located behind leaded glass walls and operated with master–slave manipulators (Fig-

Figure 9.1. AVC-80 moisture–solids analyzer.

Computer

Figure 9.2. Drying–digestive system. For chemical dissolution, drying, ashing, and sintering of materials. A programmable microprocessor digital computer permits heating or drying in three stages with separate control of microwave power and heating time. A turntable permits multiple samples to be run at the same time and a variable exhaust system allows complete control of air flow in the microwave cavity.

ure 9.3). Some instrument components, such as electric insulation and electronic components, are sensitive to radiation damage (2, 3) and cannot be used in high-radiation environments.

The instruments were studied component-by-component to see which components had to be, could be, or should be remotely operated. The components were examined for operability, reasonable service life, and accessibility for maintenance. The mechanical mountings of the nonremotely operated components on the AVC-80 and MDS-81 were simplified and modified. If these components failed or wore out, they could be replaced with the master–slave manipulators. All modifications were subject to time and cost restraints.

The following components remained with the microwave unit in the HLC: the magnetron, the magnetron filament transformer, the weighing sensor of the balance module, the wave guide and magnetron cooling fans, the exhaust fans, the wave guide and magnetron thermal switches, the door safety interlock switch, and the mode stirrer and turntable motors.

The remotely operated components were the microcomputer (the input–output sections including the digital power supply, computer cards, keyboard, and display); the analog electronics for the weighing system; and the high-voltage power supply for the magnetron. The remotely operated components were housed in a cabinet with a slide-out electronics chassis

Figure 9.3. High-level radiation cell (HLC). The operator is seen using the master–slave manipulators behind the leaded glass wall of the HLC. The manipulators move mechanical "hands", inside the cell, which mimic the motion of the operator's hands outside the cell. The remotely operated components of the AVC-80 moisture–solids analyzer with printer are shown on the left.

(Figure 9.4). The design and layout of the cabinet was accessible, and easily serviced. All indicators and circuit-protection devices were on the front panel. The MDS-81 remoted cabinet and the microwave oven are shown in Figure 9.5 before installation in the HLC.

Radiation Effects on Electronic Components

General

Nuclear radiation effects in the HLCs at the Savannah River Laboratory are produced by α-, β-, and γ-radiation. Some neutrons are generated by the reaction of α-, β-, and γ-radiation with materials in the cell, but they are of relatively low intensity and are insufficient to cause radiation damage.

Figure 9.4. Cabinet, with slide-out chassis, containing the remotely operated components of the MDS-81 drying–digestion system.

Primary α- and β-radiation produce intense ionization but do not penetrate deeply enough to be a major source of damage. γ-Radiation is the most damaging because it penetrates deeply producing β-radiation, which in turn causes intense ionization.

Organic materials in electronic components are the most susceptible to damage (2, 3). The intense ionization generated by radiation breaks the chemical bonds or linkages that bind the atoms into molecules. The fragments of the disrupted molecules react to form new compounds by cross linking, scission, and irradiation-induced oxidation.

The amount of radiation, the "dose", will be given in roentgens (R). A roentgen is the amount of ionizing radiation (γ- and X-rays) that imparts 83 ergs per gram of air. Because the type of radiation and the type of material (air) are specified, the roentgen is an indirect way to describe a radiation field comprised of γ-and X-rays. The actual amount of energy to which an electronic component is exposed depends on the type of radiation; the chemical elements present; the component's shape, volume, density, molecular arrangement; and other factors. Hence, the description of radia-

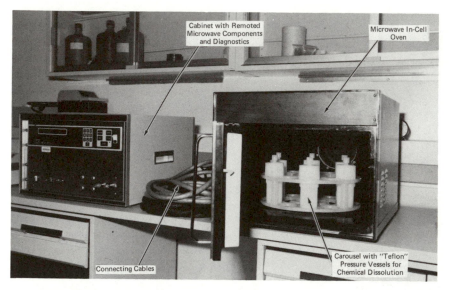

Figure 9.5. MDS-81 remotely operated microwave system.

tion damage in terms of roentgens is at best empirical. The area of the cell in which the microwave unit is located "sees" a γ-radiation field of 1000 R/h.

Estimated Life of Microwave Electronic Components

The insulating material in the magnetron and wave guide thermal switches is a high-temperature epoxy that suffers only negligible damage from γ exposures of 2.4×10^9 R (or 275 years in a field of 1000 R/h). The "Tefzel" wiring insulation will last 14 years in the same radiation field.

The magnetron is made from a copper alloy, so it is unaffected by α-, β-, or γ-radiation. The radiation might cause a few extra electrons to be freed from the cathode, but probably not enough to shorten the life of the emitter surface. The glass-to-metal seal used for isolation of electrical potentials also has great irradiation resistance. Glass dielectric capacitors have been exposed to 3.6×10^8 R and remained within their capacitance tolerance. It may be possible to have a high-voltage breakdown due to ionization of the air in the cell because there is a 4000-V potential across the leads coming out of the magnetron. However, a breakdown has not occurred. The magnetrons, however, have been in successful operation for 2 years.

The insulation for the wire windings in the motors of the mode stirrer and turntable and the cooling and exhaust fans, is a (polyvinyl)formal such as "formvar." The (polyvinyl)formals are thermoplastic resins and may experience mild radiation damage in 3 years at 1000 R/h, but are satisfactory for many applications. We have successfully operated one of the fan motors in the cell for 3 years as an advance test for the other components.

The safety-door interlock switch case is made of an unfilled phenolic resin and is mildly damaged after only 3.6×10^6 R γ exposure and moderately to severely damaged at 2.4×10^7 R or about 2.7 years at 1000 R/h. This switch is not needed for operator safety because the microwave unit is shielded from the operator by a 4-foot wall of leaded glass. This switch shuts off the magnetron. The sample turntable is polypropylene and has about the same expected life as the unfilled phenolic material in the interlock switch. It is routinely taken in and out of the unit, so it is easy to replace.

For corrosion resistance, the inside of the microwave cavity of the MDS-81 drying–digestion system is coated with Teflon. The coating is a combination of Teflon PTFE [poly(tetrafluoroethylene)] and Teflon FEP. FEP is tetrafluoroethylene–hexafluoropropylene copolymers. No data are available for Teflon PTFE but Teflon TFE is severely damaged at 6×10^4 R. Teflon FEP is more resistant but is severely damaged at 8.4×10^5 R. One would expect the coating to be severely damaged in one month in a field of 1000 R/h; however, the coating is only 0.002 in. thick and does not have to provide structural support. Furthermore, the unit is not used for open-vessel acid dissolution, so corrosion should not be a problem.

The pressure vessels for chemical dissolution are made of Teflon PFA (perfluoralkoxy), for which there is no data on the effects of radiation. However, Teflon PFA has superior mechanical properties and a higher melting point than other Teflons, so it may have somewhat improved radiation resistance. The pressure vessels, which are replaced frequently, are shielded when not in use.

Experimental Procedures

Experimental procedures were developed using a radiation-free "mock-up" cell. The procedures take more operator training time to develop because, in place of direct hand contact, master–slave manipulators are used (Figures 9.3 and 9.6). Simple operations such as transferring liquid or solid samples require the design of special tools and techniques, for example, the gripping attachments on the pipet and tweezers (Figure 9.6).

Weight Percent Solids Determination of Nuclear Waste Slurries

Procedures for the determination of weight percent solids with the remote-controlled AVC-80 moisture–solids analyzer were relatively easy to develop because of their simplicity and the degree to which the microwave unit is automated. In the microwave method, a slurry of nuclear waste is dried and weighed, and the weight percent of solids is calculated and recorded by using the built-in balance, microcomputer, and recorder. The general procedure for determining the weight percent of solids is given in the operating manual for the AVC-80.

The waste slurry is pipetted onto a glass-fiber pad, another fiber pad is placed over it to make a "sandwich", and the sandwich is gently pressed together to spread the slurry through the pads. The sandwich is then placed on the balance pan in the oven. The modified tweezers for handling the pads and the flat "egg turner" used to press the pads together are shown in Figure 9.6. The volume of the slurry does not need to be known, but it must not exceed the absorption capacity of the pads. The pads increase the surface area of the slurry to speed drying and preventing slurry from "bubbling off" during heating.

The microcomputer is programmed to weigh and record the initial weight of the slurry sample and to control the drying operation. The sample is dried for a preset time, or until the weight does not change by a preset amount (e.g., 0.002 g in 15 s). The computer then records the final dried-slurry weight and from this data calculates the weight percent of solids. For a microwave power of 25%, the drying time for waste slurries is 4 to 6 min. The power level should be kept low enough to avoid decomposition of the sample.

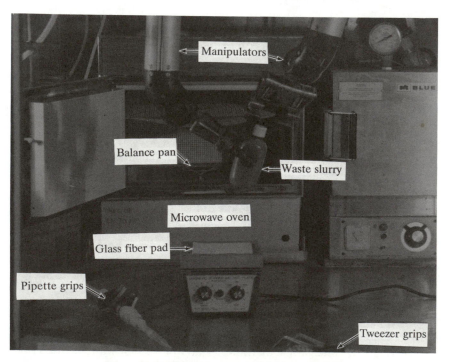

Figure 9.6. Weight percent solids determination.

The AVC-80 has an accuracy of ±0.1% for samples with loss-on-drying of 200 mg or more; however, the precision of measurement for the waste slurries is limited to a relative standard deviation of 1% because of "settling" of the slurries during sampling. A smaller source of error is the drying of the slurry on the porous glass-fiber sample pad before insertion into the microwave oven. The sample pads should be dried in the microwave oven before the sample is placed on them. Settling is minimized by vigorously shaking the sample vial with the manipulators and a Vortex mixer. To further homogenize the slurry, the sample is drawn into the pipet and expelled quickly back into the sample vial several times before drawing the sample.

Samples that settle too quickly to obtain a homogeneous sample (e.g., glass frit) require filtration to separate the solids from the liquid. The solids are then dried and weighed by the AVC-80. The total weight percent of solids can then be calculated from the dried weight divided by the total weight of the sample (i.e., the liquid plus the solids). The weight percent of soluble solids may be obtained by analyzing a portion of the filtrate by the procedure in the operating manual fro the AVC–80. The difference between the total solids and soluble solids is the insoluble solids.

Calcined Weight Percent of Solids

In the general procedure for weight percent of solids determination, the total weight (i.e., the weight of the liquid plus solids) is recorded and the sample is dried to a powder. The calcined weight percent of solids is determined by placing the dried sample in a furnace and heating it at the calcining temperature for the 30 min, and placing it back in the AVC-80 for a final weighing. This final weight divided by the initial total weight × 100 is the calcined weight percent of solids. The software of the AVC-80 will be modified, in the next unit that is remoted, so that calcined weight percent of solids can be calculated automatically, as it is for the conventional weight percent of solids. Quartz fiber pads are available and should be used for higher temperature calcining.

The accuracy of the AVC-80 may be checked with an 8 weight percent NaCl solution prepared by weighing 8.000 g of NaCl into a 125-mL plastic bottle and adding deionized water to make the final weight of the solution 100 g. The bottle should be kept sealed to avoid evaporation.

Figure 9.6 shows an operator performing a weight percent of solids analysis of a nuclear waste slurry. The analysis, which previously required 24 h to perform in the cell, now takes 15 min.

Microwave Dissolution of Simulated Nuclear Waste

Dissolution techniques for "simulated" nuclear wastes were developed at the National Bureau of Standards (NBS) in collaboration with H. M. Kingston and L. B. Jassie. The dissolutions were performed in sealed Teflon pressure vessels at elevated temperatures and pressures. The Teflon is the perfluoroalkoxy (PFA) fluorocarbon resin, called Teflon 340. Teflon 340 has mechanical properties superior to other Teflon products (i.e., tensile strength is 4000 psi at 23 °C; melting point is 302–310 °C). The vessels were obtained from CEM Corporation and have a pressure-relief valves that open at 125–150 psi. The CEM Corporation also markets a capping station that allows the vessels to be sealed to the same torque each time. The capping station was modified for remote operation by Floyd Associates.

To ensure safe operation, specific microwave power levels and heating times were developed at NBS. H. M. Kingston and L. B. Jassie (4, 5) have equipped the MDS-81 microwave digestion system to measure instantaneous temperatures and pressures during dissolution in sealed pressure vessels.

The compositions of the simulated nuclear waste and the glass frit with which it is mixed for vitrification are given in Table 9-1. The dissolutions were performed on the simulated waste alone and on a composition of 72 weight percent of glass frit plus 28 weight percent of simulated waste.

Procedure for Microwave Dissolution of Simulated Nuclear Waste

1. Place approximately 0.25 g of waste or waste plus frit in each of two 120-mL Teflon pressure vessels (duplicate samples). The weights of the samples should be recorded accurately

Table 9-1. Compositions of Simulated Nuclear Waste and Glass Frit

Simulated Waste		Glass Frit	
Compound	wt %	Compound	wt %
$Fe(OH)_3$	44.61	SiO_2	69.3
$Al(OH)_3$	18.42	Na_2O	11.8
MnO_2	8.80	B_2O_3	12.9
$CaCO_3$	6.90	Li_2O	4.16
NaOH	4.95	MgO	1.72
SiO_2	4.92	Fe	0.038
$Ni(OH)_2$	4.00	Mn	<0.005
$NaNO_3$	2.84	Ni	0.015
Zeolite	2.41	Cr	0.02
Rare Earths	0.567	Pb	0.016
CuO	0.370	F	<0.05
Cr_2O_3	0.364	Cl	0.005
KOH	0.336		
$NaAl(OH)_4$	0.253		
$Ca(PO_4)_2$	0.210		
Na_3PO_4	0.050		
Total	100.0		

because they will be needed for subsequent quantitative chemical analysis.

2. Add 5 mL of concentrated HF and 5 mL of concentrated HNO_3. It is best to prepare a specific amount of each reagent for each sample in a small plastic bottle outside the cell. **Experience with safe handling of HF and HNO_3 is required for anyone doing these dissolutions.**

3. Use the capping station to automatically screw the cap on each pressure vessel to the same torque. (The torque is set by the vendor).

4. Heat the two pressure vessels in the MDS-81 microwave oven at 50% power for 10 min on the rotating turntable in the 12-vessel holder (*see* Figure 9.5). Place the two vessels symmetrically across from each other. Make sure the turntable is rotating during heating or one of the vessels may stay in a "hot spot" and overheat. Overheating would activate the pressure relief valve and abort the experiment. Figure 9.7 shows a typical plot of temperature and pressure vs. time for dissolution of 0.25 g of waste in 5 mL of HF plus 5 mL of HNO_3 in the NBS-instrumented MDS-81 microwave system.

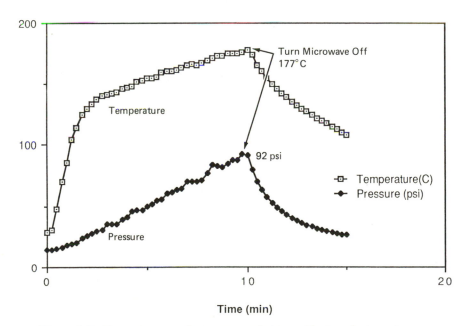

Figure 9.7. Temperature and pressure variations with time during microwave dissolution of simulated nuclear waste. The small fluctuations in temperature and pressure are real and are due to the off–on duty cycle of the magnetron.

5. Remove the vessels and cool them sufficiently so they can be handled safely. A blower can be used, but the tubes must be cooled *uniformly*, otherwise, the vessels will contract at a different rate than the caps and leak hot acid. Leaking would abort the experiment or create an unsafe condition.

6. Remove the caps from the vessels by using the capping station, then add 40 mL of heated (below boiling) deionized water containing 1.5 g of boric acid. Stir each vessel thoroughly with a Teflon or other acid-resistant stirrer, and pour each resulting solution into a 100-mL Teflon beaker. Heat on a hot plate or in a furnace at near boiling temperature for 15 min while stirring intermittently with a clean teflon stirrer. The boric acid complexes the fluoride and thereby dissolves the insoluble metal fluorides.

7. Add 5 mL of concentrated HCl to each vessel and continue to heat for 30 to 45 min until the solution is clear. Insoluble transition metal precipitates that form strong chloride complexes are dissolved in this step. The clear solution may be pink, amber, or pale green, depending on the amounts of iron and silicon, and on whether the waste was calcined before dissolution.

   NOTE: We have successfully used microwave heating for 3 to 5 min at 50% power in steps 6 and 7. However, this is more labor-intensive than the procedure just given and does not save much time because the pressure vessels have to be capped, heated, cooled, and uncapped after additions of boric acid and HCl. We selected 5 mL of HF plus 5 mL of HNO_3 because we had to cover a range of materials with widely different compositions. For waste without glass frit, 3 mL of HF and 7 mL of HNO_3 should be used because the waste it contains more iron and less silicon. The data shown in Figure 9.7 and in Table 9-1 were for simulated nuclear waste. However, we have successfully dissolved actual nuclear waste.

Electronic Modification of the AVC-80 Moisture–Solids Analyzer

Wiring.   All wiring in the remote-operated section of the microwave units was insulated with Tefzel for irradiation resistance. Extra wire pairs were included in the cables in case additional wiring was necessary or wiring failed in the cell. Special indicating fuse holders were used along with other diagnostic indicators to allow early detection and diagnosis of cell-component failures or remote electronics problems.

Cable.   Long cables were used to connect the cabinet containing the remote components and the microwave oven in the cell. Cables at least 16 ft long, and preferably 26 ft long, give flexibility in locating the equipment inside and outside the cell. Three cables were used: one for the analog weight sensor, a second for the AC power wiring, and a third for the high voltage to the magnetron. Careful grounding of each cable, each cable pair, and interconnecting grounds provided safety, noise immunity, and radiation resistance.

The Weighing Sensor.   The electronic weighing system is very sensitive because noise pickup in long cables affects low-level analog signals. There-fore, a separate shielded cable with individually shielded pairs of wires was used to connect the remote weighing-sensor to the electronics outside the cell. This arrangement permitted the weighing sensor precision to be main-tained at ± 0.1 mg. The weighing sensor is enclosed in a metal housing, so it is shielded from radiation by both this housing and the walls of the lower cabinet of the microwave unit.

Microcomputer and Input–Output Sections.   All of these compo-nents, including the digital power supply, computer cards, keyboard, and display were incorporated into the remote cabinet.

Microwave Power System.   The microwave generator (magnetron tube) and its filament transformer remained with the microwave oven. The microwave power supply was housed in the remote electronics cabinet. Special high-voltage connectors and cabling were used to connect the high-voltage supply to the magnetron in the cell.

Mechanical Modification of the AVC-80 Moisture–Solids Analyzer

Microwave Cover.   The front edges on the cover were shortened so the cover did not have to slide into the grooves on the front of the upper microwave cabinet. Guide plates were added to both sides of the unit for alignment. The cover is fastened to the back of the cabinet with only two screws. The three screws that are normally used to fasten the cover to each side of the cabinet were eliminated.

Microwave Door.   Hinge modifications were made to let the door open wider. Spring tension on the door latch was reduced to make the door easier to open. The safety interlock switch at the door hinge was removed. The safety interlock switch at the door latch was retained, but its mounting bracket was modified so that it mounts on studs, with ¼-turn "pal" nuts, behind the door seal. Hence, if this switch fails it can be easily replaced with a nut driver. Only one safety switch was necessary (for out-of-cell checkout of the unit) because the microwave will be located behind a 4-foot-thick leaded glass wall.

Cavity Exhaust Fan and Wave Guide Cooling Fan. The fans were modified so they could be pressed onto four mounting studs and locked into place with four ¼-turn pal nuts. The magnetron cooling fan is of the same design as the other fans but is attached to a piece of venting duct. The magnetron fan and duct are removed as a single assembly by removing the three screws that attach the assembly to the top of the oven cavity.

Balance Module. The entire balance module was modified to attach to the floor of the oven cavity in the upper cabinet of the microwave unit. In the AVC–80 the radio frequency (rf) stub was attached to the oven cavity floor and the balance pan was attached to the lower base cabinet. Hence, after the upper cabinet was raised for service, the cabinet had to be lowered in a precise manner so the balance stem passed through the center of the rf stub. For easy replacement, an adapter flange was installed on the balance module so that it could be screwed directly to the rf stub. The threads on the rf stub and the balance module are self-centering, so alignment of the balance stem into the center of the stub is unnecessary.

Balance Air Shield. The air shield around the balance pan was enlarged and the front face removed to make it easier for the manipulators to place the sample pads and sample on the balance pan. The front face of the shield was unnecessary because the shield fits close to the door of the microwave oven.

Mode Stirrer. The mode stirrer motor is mounted on two studs and fixed with ¼-turn pal nuts on top of the oven cavity. The mode stirrer blade mounting shaft was modified to eliminate the existing retainer-pin design. The end of the shaft was flared so the blade was held on the shaft by a pressed-on retainer washer.

Magnetron. The spring clips on the cabinet mounting bracket were loosened so that it was easier to slip the ears of the magnetron mount under them.

Magnetron Thermal Switch. The magnetron thermal switch mounting was relocated to make its spring clip accessible to the manipulators.

Magnetron Filament Transformer. The transformer was relocated to the lower cabinet for easier accessibility.

Electrical and Mechanical Modifications to the MDS-81 Drying–Digestion System

The modifications were the same for the MDS-81 as for the AVC-80, with the following exceptions:

- The MDS-81 does not have a balance module.

- The MDS-81 has a large cavity blower and outlet for exhausting acid fumes.

- The MDS-81 has a motor-driven turntable that rotates samples during microwave operation to reduce the effects of hot spots.

If the exhaust blower failed the cavity would not be protected from acid exposure, and the temperature inside the pressure vessels would increase. Therefore, a safety interlock circuit was installed to shut off the magnetron if the blower failed.

The exhaust blower was made easier to replace by attaching eight studs to the blower housing face for alignment of the blower and to hold it in place when nuts are being attached from inside the oven cavity. To replace the blower, one has to first remove the wave guide cooling duct. The duct is attached to the top of the oven cavity by three screws. This duct is also removed to replace the wave guide cooling fan.

In the unmodified oven, the turntable motor is inaccessible to the manipulators. Therefore, it was mounted in a separate metal box and attached to the back of the microwave cabinet. The turntable motion is visible to the operator so the microwave power can be shut off if the table stops turning.

Conclusions

The remote-operated AVC-80 and MDS-81 microwave systems are well suited to HLC operation. The systems have been operating for 2 years with only minor maintenance.

The remaining components could be modified for remote operation (e.g., the magnetron could be operated remotely and the microwaves piped into the cell, or the fans could be replaced by forced air from the outside). However, the cost would be many times the cost of the present system and does not appear to be justified in this application.

The radiation resistance of the motors and switches left in the cell could be improved by building them with Tefzel insulation. However, the "off-shelf" components used are inexpensive, mounted for easy replacement, and are estimated to last at least 3 years in a γ-radiation field of 1000 R/h.

Experimental procedures using microwaves save a great deal of time compared to conventional procedures when operating in or outside of a high-level radiation cell. The procedure for chemical dissolution of wastes and waste glasses may be further optimized if the range of compositions encountered is less varied. Safe operating power levels and heating times must be established before any work is done.

Remotely operated microwave systems may also be useful in chemical hoods or in other facilities where equipment needs to be protected from corrosion or the operators need to be protected from the environment. Sealed-vessel microwave dissolution techniques may be the wave of the future, not only because they are cheaper and faster, but because acid and other toxic vapors are not vented into the environment.

Acknowledgments

The authors thank the following for their contributions. C. E. Whitney of Savannah River Laboratories, who has long experience and competence in HLC work, guided us in the design and the development of in-cell experimental procedures. M. J. Collins and R. J. Goetchius of the CEM Corporation provided assistance and encouragement in this microwave application and development. R. W. Lewis, Engineering Physics Laboratory, E. I. du Pont de Nemours and Company, provided helpful discussions on methods of remotely operating the magnetron and possible operating characteristics of a magnetron in a nuclear radiation field. The information contained in this article was developed during the course of work under Contract No. DE–AC09–76SR00001 with the U.S. Department of Energy.

Literature Cited

1. Baxter, R. G. *Design and Construction of the Defense Waste Processing Facility Project at the Savannah River Plant*, Proceedings of Waste Management 1986 Symposium, Tucson, AZ, March 3–16, Waste Isolation in the U.S.; 1986, Vol. 2, pp 449–454.
2. DPE–3586, *Radiation Effects on Nonmetallic Materials and Components*, Savannah River Plant Work Request 860755 Project S–1780, compiled by E. I. du Pont de Nemours and Company, Inc. Engineering Department, Wilmington, Delaware, September 1978.
3. Thatcher, R. K.; Hamman, D. J.; Chapin, W. E.; Hanks, C. L.; Wyler, E. N. The Effect of Nuclear Radiation on Electronic Components, Including Semiconductors, Battelle Memorial Institute Radiation Effects Center, Report REIC-36, October 1, 1964.
4. Jassie, J. B.; Kingston, H. M. Microwave Dissolution in Closed Vessels Under Elevated Temperature and Pressure, Pittsburgh Conference Abstracts, 1985, Paper No. 108A.
5. Kingston, H. M.; Jassie, L. B. *Anal. Chem.* **1986**, 58, 2534–41.

Received for review April 10, 1987. Accepted revised manuscript January 22, 1988.

CHAPTER 10

Manual and Robotically Controlled Microwave Pressure Dissolution of Minerals

John M. Labrecque

"The principal mark of genius is not perfection but originality, the opening of new frontiers".

Arthur Koestler

A manual microwave pressure digestion system is described to provide rapid, complete, and contamination-free dissolution of powdered samples. Laboratory safety practices from the development stage and current safety procedures are reviewed and recommended. Also discussed in this chapter is the installation of a Zymate II robot system, which serviced several modules: balance, acids dispenser, microwave heating unit, and an automated capper, on a 125- × 240-cm (4- × 8-ft) area. It achieved the goal of performing, in its entirety, microwave pressure dissolution. The powdered sample was weighed, dissolved with mineral acids (HNO$_3$ and HCl) in a laboratory microwave heating unit, and made up to final volume by flexible automation. Productivity, reliability, and performance of the commercial system are assessed.

THE MINING AND METALLURGICAL PROCESSES for base and precious metals at Kidd Creek Mines are guided by 750,000 analyses per year conducted by fire assay (FA), X-ray fluorescence (XRF), optical emission spectroscopy (OES), atomic absorption spectroscopy (AAS), and inductively coupled plasma emission spectroscopy (ICP). Lengthy digestion with nitric and sulfuric acids in open beaker is necessary to prepare mineral samples for an AAS or ICP nebulization system.

The dissolution of an ore sample using microwave technology has been a significant development. The last 20 years have witnessed a steady improvement in analytical instruments. Despite these advances, the traditional sample dissolution procedure was still in use. The publication of *A Microwave*

1450–6/88/0203$08.00/0

System for the Acid Dissolution of Metal and Mineral Samples by Matthes et al. in 1983 (*1*) finally changed that trend. For the first time, a copper smelting slag sample was rapidly dissolved, analyzed for silica without a NaOH matrix, and reported to production within the hour. The original paper recommended polycarbonate bottles for the microwave dissolution. However, leakage caused by increased internal pressure in the polycarbonate bottles during this type of dissolution was causing severe problems in our laboratory, and the technique seemed uncontrollable. John Bozic from Inco (Sudbury, Ontario) suggested the use of a Teflon [poly(tetrafluoroethylene)] PFA digestion vessel manufactured by the Savillex Corporation. The Teflon dissolution vessel is capable of safely sustaining internal pressure up to 520 kPa (75 psi) above ambient room temperature and internal temperature reaching 127 °C, according to the manufacturer. The Teflon PFA digestion vessel has performed consistently and reliably.

The process of capping and uncapping the vessel is tedious. About seven to eight turns are necessary to complete the capping of the vessel to 16–19 N·m (12 to 14 ft-lb) of torque. A robotic system has been developed to speed this process and decrease the workload of sample digestion.

Distribution of Selenium

Selenium, a minor constituent of the unusual Kidd orebody, has a direct effect on quality of the copper produced by the pyrometallurgical smelting and tankhouse refining processes. To improve the quality of our product, a study was undertaken to monitor the distribution of selenium in the Kidd orebody, specifically in the bornite zone. The Kidd bornite contains up to 60% copper and contains the following impurities: selenium, silver, cobalt, arsenic, and bismuth. Original wet chemical assays compared to neutron activation analysis showed that only 70% of the selenium was recovered. Early comparisons showed that a dissolution with nitric and hydrochloric acid under pressure by microwave energy resulted in recoveries greater than 90%. Our first objective, which was the delineation of selenium in our orebody, was successfully accomplished with manual microwave pressure dissolution and Zeeman flame atomic absorption analysis (ZFAAS).

Problems Associated with Manual Pressure Dissolution

Safety is a prime concern when using acids under pressure, and a safe approach to the pressure dissolution by microwave energy was essential. Three guidelines were established for using the Teflon PFA pressure dissolution vessels in the microwave heating unit:

- Never exceed 300 mg of sulfide ores or concentrates.
- Let the reaction of about 10 mL of mineral acids subside before capping the vessel.
- Do not to exceed 30 s per sample at 700 W in the microwave oven.

These guidelines were derived from the U.S. Bureau of Mines Technical Progress Report (1) and our own observations.

Microwave Heating Unit Location

Microwave heating unit exposed to acids and operating in a fume hood are subject to premature corrosion and electronic failures. It is strongly recommended that a microwave heating unit not be located in the fume hood. The ideal location is adjacent to the hood and exhausting into it.

Digestion Vessel Safety Inspection

The practice of closely inspecting the vessels for flaws or fissures was incorporated into our laboratory routine. The life span of the vessel and the warning signs indicating the end of the useful life span are unknown. The current vessel life span is from 4 to 6 months. Our practice is to replace a vessel when the original purchase cost is recovered.

Weight of Sample To Be Dissolved

When 90% of the sample passed through 44-μm (325-mesh) fine screen, it had been traditional to use a sample weight of 1 g for ores and 0.5 g for concentrates to start a dissolution. Microwave digestion does not allow these weights when a complete dissolution under normal temperature and pressure is required. An evaluation must be made to determine a suitable sample weight. To illustrate this concept, Canmet standard reference material (SRM) SL-1, blast furnace slag was analyzed for silica by ICP with consistent digestion parameters. Figure 10.1 demonstrates the suitable sample weight derived from a plot of the variance of chemical analysis obtained from 10 analyses against the sample mass used. The least variance area statistically shows the preferable weight for chemical digestion (2).

Capping and Uncapping the Vessel

High sample throughput using microwave dissolution became a tedious procedure, and alternatives were sought to replace manual capping. A trial run

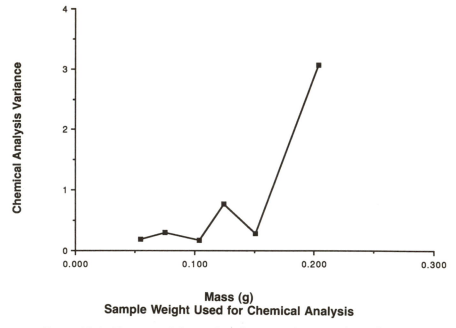

Figure 10.1. Variance of chemical analysis as a function of sample mass.

of an automatic capping device was not successful, but the concept of completely automating with flexible automation was first envisioned. The introduction of an air-driven torque wrench did speed up the capping and uncapping process, but virtually transformed the laboratory into a machine shop. Noise levels of up to 90 dB were recorded at the capping station, and the use of hearing protection became mandatory (3).

Need for Flexible Automation

Reducing the cost per determination is a major factor in seeking automation. Following the Pittsburgh Conference of 1985, the possibilities for automation became more evident with the technological advances made at the Zymark Corporation.

Automation requires more training time to keep abreast of modern technology. Meanwhile, the analyst must maintain productivity or sample throughput. Therefore, an automated microwave pressure dissolution system must be

1. Easy to operate

2. Autonomous (operates completely independently)

3. Reliable (operates unattended and provides uniform application)

With these objectives attained, automation can increase productivity and provide the analyst with time to discover, understand, and implement the full potential of modern instrumentation in the laboratory.

Technical Innovations

Three major advances in the field of analytical chemistry coincided to make robotically controlled microwave pressure dissolution possible:

1. Better recovery rates with microwave dissolution than with the open-beaker method.
2. Use of Teflon PFA pressure dissolution vessels.
3. Availability of advanced flexible automation in robotics.

Microwave Pressure Dissolution with Proven Chemistry

The microwave dissolution method yields significantly higher recovery rates for selenium than the traditional open-beaker method. During the study on the distribution of selenium in the Kidd sulfide orebody, recoveries of over 91% were recorded by manual microwave pressure dissolution compared to 70% achieved by the wet chemical method. These recovery figures were corroborated by neutron activation analysis that had a coefficient of variation (CV) of 8%. Verification of the analytical data and method was conducted according to the guidelines established by Canmet (4). Table 10-1 shows comparative data for selenium.

The open-beaker procedure consisted of a slow nitric and perchloric acid dissolution at low heat, followed by an extraction in toluene containing 1% by weight of *o*-phenylenediamine. Absorbance of the toluene solution after extraction was measured at 335 μm versus a chemical blank. A suitable calibration was used to determine selenium from the absorbance value. This labor-intensive method required two analysts to prepare 50 samples in order to have results reported the following day.

The following procedure was used for the microwave dissolution of a tray of 18 samples. The sample (300 mg) was weighed directly into each vessel, 3 mL of concentrated nitric acid was added, and the mixture was allowed to react for 2 min. Each vessel was sealed at 16–19 N·m (12–14 ft-lb) torque and heated for 3 min at 280 W. The vessel was cooled in a water bath and uncapped, and 5 mL of concentrated hydrochloric acid was

Table 10-1. Comparison of Selenium Results from Ore Samples

Sample No.	NAA[a]	Colorimetri[b]	ZFAAS[c]
6495	180	126	142
7011	320	224	311
7372	80	56	66
7397	90	63	74
7733	67	47	51
7740	170	120	147
7744	1550	1085	1551
8019	140	98	124
8430	934	654	985
9982	2700	1890	2723

NOTE: All results are given in micrograms per gram.
[a]Neutron activation analysis.
[b]Acid dissolution in HNO_3–$HClO_4$ followed by extraction and colorimetric determination.
[c]Acid dissolution HNO_3–HCl using microwave heating under pressure and Zeeman flame atomic absorption analysis.

added. The vessel was capped again and heated for 3 min at 280 W. The vessel was cooled again and uncapped, and 22 mL of water was added. The complete tray was recapped and microwaved for 5 min at 600 W. Because there was no volume is loss by evaporation, the vessel was cooled, mixed, and uncapped, and the contents were poured into a flint glass test tube for analysis. After implementation of microwave oven dissolution followed by ZFAAS, one analyst can produce 50 samples for the results the same day. Other elements such as arsenic, copper, and zinc can be determined from the same dissolved sample.

Teflon PFA Pressure Dissolution Vessels

In the past, the weak link in microwave applications had been the dissolution vessel. The Savillex Corporation design of the Teflon PFA digestion vessel and its availability at an acceptable cost made microwave dissolution possible in our laboratory. An important aspect of the vessel design is that it is easily handled by the robot's general-purpose fingers. The newly purchased vessels should be annealed in a convection oven for 5 working days at 110 °C to improve strength, reliability, and life span.

Advanced Flexible Automation in Robotics

The complexity of executing the sample weighing and digestion steps demanded a programmable flexible automation approach rather than a fixed approach as used in the manufacturing industry. During the summer of 1985 the only system capable of performing complex manipulations and dispensing a powdered sample to a targeted weight was the Zymate system. Other advantages of the Zymate system were that the manufacturer had already initiated interfacing between the controller, the balance, the microwave heating unit, and the modified vessel capping device.

Equipment Used for Manual Microwave Dissolution

The Microwave Heating Unit

The microwave unit was a CEM Corporation MDS 81-D. A laboratory microwave heating unit must incorporate certain features to achieve safe and reproducible results. A tight-closing door with a reliable power interruption device for accidental door opening, is a most important safety feature, along with a reliable and variable-speed control exhaust system. The savings realized by purchasing a kitchen microwave oven are offset by poor reliability and inconsistent results. Only a laboratory model will provide consistent and reliable results. The power available (P_{abs}) from the microwave unit is detectable and can be determined by using the following method recommended by the manufacturer.

$$P_{abs} = 35(T_f - T_i) \tag{10.1}$$

The temperature of 1 L of water is measured initially to the nearest 0.1 °C (T_i). The container of water is placed in the microwave unit (with exhaust off) and 100% power is applied for exactly 120 s. Immediately, another temperature measurement is taken (T_f), and the increase in temperature is multiplied by 35. The average power available should be between 570 and 630 W. This verification is conducted in our laboratory weekly to monitor the magnetron efficiency.

Pressure Dissolution Vessel

Teflon PFA vessels with built-in safety relief valves are available from the CEM Corporation. These safety valves will open repeatedly if pressure inside the vessel exceeds 830 kPa (120 psi). Once the sample's reactions are predictable, and high throughput is required, use of the Savillex vessel with no vent in the cap is very successful for sulfide ores, concentrates, slags, and metallurgical byproducts.

Trays for the Teflon PFA Vessels

Trays to hold the digestion vessels during microwave heating were constructed according to the guidelines given in the U.S. Bureau of Mines

Report (1). To make certain that a cap cannot fly off during of after routine dissolution, a 5-mm thick polypropylene circle is inserted into the T-shape handle and turned 90°. This safety precaution ensures that if a cap is cross threaded or fails under pressure it will not cause injuries to the analyst.

Capping and Uncapping the Teflon PFA Vessels

Currently, a semiautomatic device is available to accomplish this task. To speed up the work in our laboratory, the vessel capping station is equipped with an air-driven torque wrench and a socket. To hold the vessel, the lugs on the body of the dissolution vessel are held securely in a 25-mm thick polyethylene socket base held about 10 cm above the station bench. Excessive tightening of the vessel cap will cause acid fume leakages. To ensure proper capping, the air-line pressure is regulated at 390 kPa (57 psi).

Equipment Used for Robotic Microwave Dissolution

Currently, the system operates on a 122- × 244-cm (4- × 8-ft) surface that has one robot arm, an analytical balance, an acid dispensing system, a capping device, a microwave heating unit and three acid-fume exhausts. Figure 10.2 gives an overall view of the system and Table 10-2 gives a description of the equipment.

Programming and Running the System

The computer (the controller) was programmed by using the topdown approach (5). A very clear understanding of the overall objective is necessary for a precise plan to be conceived and executed.

The method was divided into nine laboratory units of operations (LUOs), and each LUO was broken down to its simplest action or function (6). Absolute and relative positions were targeted with the teaching module, which simplified the programming by directing the arm, hand, or finger precisely to the selected point. Positions are called "absolute" when they are permanent locations and "relative" when they are in relation to other positions.

Robot Chemical Dissolution Sequences

Table 10-3 gives the plan that was followed to create the robotically controlled microwave pressure dissolution system.

Sequential Programming

In a sequential program, the controller executes only one task at a time. Each LUO can be observed, corrected, and fine tuned. (One sample must be processed successfully through the LUO before it can be incorporated into the main program.) Initially, our program took 27 min to complete. However, judicious programming of the controller reduced the time to 24 min.

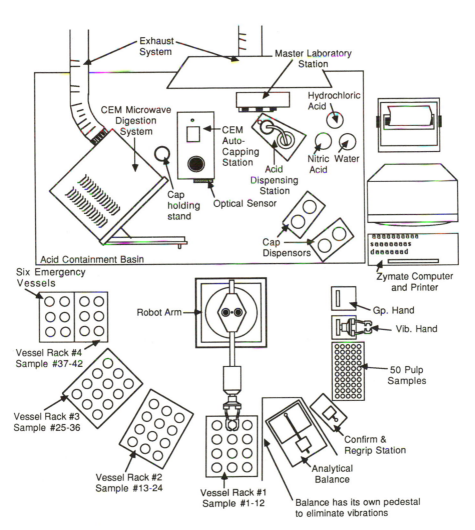

Figure 10.2. System overview.

Table 10-2. Description of Equipment Used

Equipment Used	Function
Robot arm	Flexible manipulations
Pulp sample rack	Holds 50 powdered samples for dissolution
Teflon PFA holding racks	Four racks holding 12 vessels each
Confirm and Regrip station	Confirms test tube in hand and regrips at bottom
Mettler analytical balance model AE160, 011 data output interface	Provides sample weight
Vibrating hand	Dispense sample to weight
General purpose hand	Manipulates PFA vessels and caps
Cap dispenser 1	Dispenses caps 1–25
Cap dispenser 2	Dispenses caps 26–48
Acid dispensing station	Vented location where acids and water are added
Optical sensor Reflex and proximity control (response time: 1 ms)	Confirms the presence of the vessel and cap before acid addition and capping
Capper (CEM Corp. design, Zymark modified)	Caps and uncaps vessels at 19 N·m
Microwave heating unit (CEM model MDS 81-D), three programmable stages	Microwave heating for chemical dissolution
MLS station	Prefills syringes A, B, C, and dispenses acids and water
Polypropylene safety containers	Nitric acid storage
Polypropylene safety containers	Hydrochloric acid storage
Polypropylene safety containers	Distilled water storage
Microwave door stop	Prevents microwave door from bouncing after it is opened
Two power and event controllers	Sends and receives electrical input and output to control peripherals
Computer, 179-Kb of memory 5.25-in. floppy disk	Easylab (Zymark)
Printer	Prints sample number and weight
Exhaust system	Three acid fume extractors (Lab Safety Supplies, Janesville, WI)
Peristaltic pump	Dispenses hydrofluoric acid
Utilities	Compressed air and argon, UPS (uninterruptible power supply), voltage regulator
Humidity controller	Provides consistent dispense sample to weight routine

Table 10-3. Robot Chemical Dissolution Sequences

LUO[a]	Function or Definition	Example
1. Setup balance	Preliminary prep. to tare and weigh sample	Pick up hand and place vessel in balance pedestal
2. Weigh sample	Dispense sample to weight	Seek stable weight of 300 mg, record sample number and weight
3. Add acids	Dispense 3 mL of HNO_3 and 5 mL of HCl	Move vessel to acid dispenser
4. Cap vessel to torque	Seal vessel at 19 N·m	Running torque = torque
5. Microwave sample	Dissolve sample	Micro power on or off
6. Uncap vessel	Unseal vessel	Unscrew cap counter clock
7. Add water	Dispense 22 mL H_2O	SYR.C volume = 0
8. Partially cap vessel	Replace cap on vessel	Return cap on vessel to avoid acid fumes
9. Return vessel to original position	End of cycle	Sample ready for analysis

[a] Laboratory unit of operation.

Serialization with Robotics

The serialized program permitted operation of the system because the controller managed several tasks simultaneously. The Zymate II has eight timers so that up to eight tasks can be conducted concurrently. In a serialized mode of operation, each sample is processed sequentially through the complete method, because the robot simultaneously processes several samples at different LUOs. Serialization has three major benefits over the sequential mode:

1. Efficient use of peripherals
2. Identical processing for each sample
3. Faster completion time for each sample

Efficient Use of Peripherals

This system has four principal peripheral units: (1) the balance, (2) the master laboratory station that dispenses all fluids, (3) the capper that seals the Teflon PFA vessel to torque, and (4) the microwave heating unit. The analytical balance, which is mounted on its own vibration-free pedestal, has two-way communication with the controller and provides two types of reading to the controller. One reading is an unstable weight, that is, the increasing weight in the vessel, and the second type is the final and stable weight that is printed out. The master laboratory station accurately dispenses concentrated nitric and hydrochloric acids into the vessel. The capper seals the Teflon PFA vessel to 19 N·m (14 ft-lb) torque and can unseal them as well. The torque can be adjusted by modifying a statement in the cap-to-

torque program. The microwave heating unit has its own actuated door program that is useful in the time control of the LUO. Increases in the duty cycle time of any peripheral units could mean overheating a motor or maximum use followed by lengthy periods of inactivity. The serialization mode uses all peripherals for short periods of time continuously throughout the sample dissolution cycle.

Identical Processing for Each Sample

Each suite or group of samples requires its own dissolution procedure. All geological samples are dissolved according to the established procedure, and all refinery tankhouse anode slimes are digested by a different procedure. The chemistry of this particular example of geological samples is covered in Chapter 3. For metallurgical samples, a 1-g potassium chlorate pellet is added to the vessel as an oxidizing agent during the check-list review. The potassium chlorate tablets do not react as violently as granular potassium chlorate, and the benefits of this oxidizing agent are excellent. Anode slimes are digested with 10 mL of nitric acid for 30 s at 100% power. This simple procedure in a chloride-free environment permits the titration of silver.

Faster Completion Time for Each Sample

In the sequential mode of operation, one sample takes 24 min compared to the serialized mode that completes one sample every 10 min. This time savings is directly attributable to the serialization approach. People prefer to work with a suite of samples and process them all step-by-step, but the robot–computer system excels at performing concurrent tasks. It can manage up to eight tasks concurrently.

Special Programs

Several specialized programs were created to ensure proper functioning of the system and safety of the analyst and of the system.

Initialization Program

To verify that all initial program values are properly set when the system is activated, an initialization program was created to reset 30 values. The best method of resetting is to state at the beginning of the initialization program that the robot wrist is set at zero angle, the vibrating amplitude is set at zero, and the sample in the microwave is equal to zero.

Confirm and Regrip Program

This program deals with two situations. Because there is no feedback to the Zymate controller, the confirm station microswitch acknowledges the presence of the test tube in the vibrating hand before starting the weighing routine. The second function ensures that the test tube is regripped 50 mm from the bottom and allows better dispensing of the powdered sample.

Dispense Sample-to-Weight Program

The Zymate fingers cannot manipulate a paper bag containing the powdered sample to be weighed. To overcome this limitation, the powdered sample is placed in an 18- × 150-mm test tube to a depth of 60 mm. With the help of the vibrating hand, 300 mg of the sample can be weighed directly into the vessel on the balance pedestal. The best version of the dispense sample-to-weight program is the use of the statement "IF UNSTABLE WEIGHT < 0.1 G THEN VIBRATIONS = 1" (maximum amplitude) that directly relates increasing unstable weight to intensity or pulsation of the vibrating hand. Appendix A gives a good example of this statement.

Power Failure Program

This program is activated when a power failure occurs. The uninterruptible power supply (UPS) is interrogated on its status. If the power is supplied from the UPS then the abort program is activated. During power failure, the potential for an unsafe condition can be created if a vessel is left uncapped and dissipating fuming acids over the system. Therefore, the program immediately causes a cover to be placed on the vessel and it is moved to a holding position away from the electronics. Acids are syringed back to their reservoirs, and the robot arm is moved to a safe position. The system can operate for 10–15 min before the UPS is exhausted. The system has the ability to reactivate; however, it is safer to visually verify the system's condition and check status of the support systems before reactivation. Appendix B demonstrates the approach taken to deal with power failure. Compressed-air supply and exhaust fans can become inoperative for reasons other than a power failure. The status of these fans is not currently checked by the controller.

Safety in Programming

Appendix C shows the programs used. Four systems are used to ensure safe execution of the operation: switches, optical sensors, flags, and tactile sensors.

- A safety microswitch confirms that the balance door is open or closed.

- An optical sensor confirms that a vessel is present before acid is added and that the cap is present before capping and microwave heating.

- Flags within the software confirm that the microwave unit door is open or closed. Flags are inserted in the software to confirm that an action has taken place. If the action did not occur, the command is reissued until the total amount of flags programmed is reached.

- The controller interrogates the gripforce used for tactile sensing information. If the force used exceeds the designated value then the operation is terminated before causing damage to the equipment.

System Assumptions

The system assumes

1. that the analyst has loaded approximately 60 mm of the powdered sample in to 18- × 150-mm test tubes and has placed them in order in the sample tube rack. This task is scheduled to be implemented at the sample preparation stage.

2. a sufficient number of vessels and caps.

3. sufficient nitric acid, hydrochloric acid, and water present to complete the run.

4. the three exhaust fans are operating and a supply of pressurized air and purging argon.

Self Monitoring

The controller monitors the effects of compressed gases that are used to activate doors or to purge sensitive or delicate integrated circuit boards in the robot hand. A strategically located microswitch monitors the action of door opening and reports to the controller. If the switch fails to report the door opening, the system will start monitoring flags in the program. If the door is not opened or shut then the system aborts.

Operation

The intervention of the analyst is required at the beginning of each suite of samples to be dissolved. The maximum number of samples that can be dissolved is 42. The total number of vessels present in the system is 48. Six vessels are retained in the repeat rack in case overdispensing occurs. This safety program allows the sample to be reweighed and introduced in its original position. Once an excess of sample is in the vessel, it cannot be removed! The overweight sample and vessel are disposed of at a dump site.

The analyst switches the controller to direct control and types RMD for robot microwave dissolution. The analyst is then prompted to enter the number of samples to be dissolved. The start-up is initialized by typing MD, for microwave dissolution. The time required to start-up is about 10 min, from that point on all operations are robotically controlled. To avoid any possible error, the following checklist is used before start-up.

1. Total samples to be analyzed
2. Total vessels and caps present
3. Potassium chlorate tablet requirements
4. Nitric acid, hydrochloric acid, and water levels in reservoirs
5. Compressed air valve on
6. Argon valve on
7. Exhaust systems and the main fan on
8. Syringes prefilled with acids
9. Dump site clean
10. Warmup program run to verify that all doors and fluid dispensers are functioning, and that the capper runs

Results

Dynamics of Temperatures and Pressures Within the Vessel

Information about the temperature and pressure that are developed inside a vessel during dissolution is critical to the safe operation of the system. This information is obtained by monitoring, such as that described in Chapter 6. Figure 10.3 shows the temperature and pressure reached when 18 vessels containing 300 mg of zinc sulfide concentrate in 8 mL of aqua regia are heated in the microwave heating unit for 3 min at 61% power. These parameters are for the determination of selenium, copper, zinc, and iron.

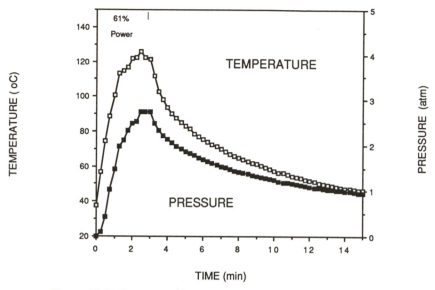

Figure 10.3. Pressure and temperature profile of zinc concentrate.

Maximum temperature of 130 °C and internal pressures of 460 kPa (67 psi) above ambient are reached. This means that pressure developed during acid dissolution is below 520 kPa, the maximum recommended by Savillex for its vessels.

Quality Control Analytical Results

Circuit samples were analyzed for a working month; the results for silver and selenium are shown in Figure 10.4 and 10.5. AAS analysis results for these circuit samples are well-documented. In our laboratory, the silver value was set at 42 ± 5.4 g/Mg. (The unit megagram is used to avoid confusion in reporting precious metals.) Selenium was set at at 507 ± 35 μg/g at a 95% confidence interval.

Analytical Results from Certified Materials After Microwave Digestion

Results of elemental analysis of certified materials digested using microwave energy are compared to data from accepted chemical dissolution in Table 10-4. These microwave digestion procedures are reported in Chapter 3.

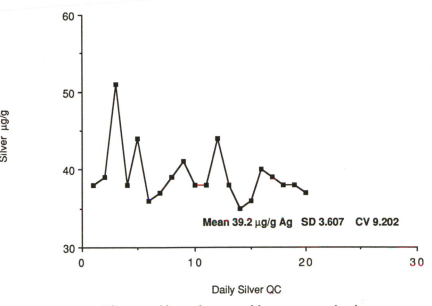

Figure 10.4. *Silver monthly quality control by microwave dissolution.*

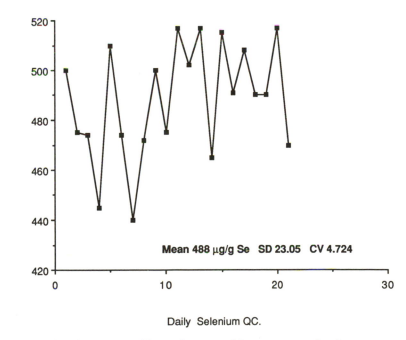

Figure 10.5. *Selenium monthly quality control by microwave dissolution.*

Silver Digestion in Chloride-Free Environment

An investigation of the precision and accuracy of our silver method revealed that dissolving our tankhouse slime by microwave heating and titrating with sodium chloride is far superior to our traditional fire assay (pyrometallurgical analytical procedure). Results for multiple determinations of the same sample are shown in Table 10-5.

The microwave–titration results, compare very closely with the Silvone gravimetric results which we believe are the most accurate (7). Microwave digestion in a chloride-free atmosphere is a rapid way to confirm the true silver value in place of the Silvone method. Silver results from the microwave

Table 10-4. Comparison Table for Manual Dissolutions

Material Description	SiO$_2$%	Al$_2$O$_3$%	CaO%
Canmet blast furnace slag SL–1	36.2 (35.7)	9.63 (9.63)	37.6 (37.5)
Canmet Cu concentrate (CCU–1)	2.90 (2.61)	0.25 (0.25)	0.10 (0.09)
Kidd internal ISRM SiO$_2$ flux	81.0 (80.0)		
Iron ore, ISRM	5.98 (5.88)		

NOTE: The values in parentheses are the certified or agreed values.

Table 10-5. Comparison of Silver Results from Digestion of Tankhouse Slime

Sample Number	Fire Assay a	Fire Assay b	Microwave–Titration	Silvone–Gravimetric Method
1	23.03	22.91	24.49	23.93
2	22.82	22.97	24.43	24.21
3	23.11	22.78	24.63	24.59
4	23.10	22.99	24.52	24.31
5	23.05	23.07	24.30	24.29
6	23.23	22.80	24.32	24.52
7	23.07	22.86	24.34	23.79
8	22.93	22.85	24.18	25.23
9	23.14	23.12	24.44	23.13
10	23.36	23.04	24.40	25.38
11	23.50	22.96	24.40	24.53
12	22.90	23.05	24.56	24.55
Mean	23.11	22.95	24.418	24.372
V	0.034	0.012	0.015	0.362
SD	0.184	0.111	0.124	0.602
CV	0.796	0.482	0.506	2.47

NOTES: All values are reported in percent. Columns entitled fire assay a and fire assay b are proof-corrected replicate silver results that agree when using different fluxes.
V is variance. SD is standard deviation. CV is coefficient of variation.

digestion–titration method are 1.4% higher than fire assay results. Such agreement is excellent.

Cost per Assay

The minimum cost for a silver determination by fire assay for a shipment of tankhouse slimes is $130. This is the cost for eight fusions (and their slags) to be proof-corrected and parted to remove gold, platinum, and palladium. On the other hand, the microwave dissolution and titration done in triplicate costs $17.25.

Robot Dissolution Results

The reliability and validation phase for the dispense-to-weight program used with our finely pulverized samples program is ongoing. The vibrating hand of the robot arm has a pulsing mechanism that shakes the sample out of the test tube. It was anticipated that this action would cause the heavy metals to sink to the bottom of the gangue. We have not observed segregation of the sample. Twenty dissolutions of the same sample were conducted, and results are shown in Table 10-6, suggest that the material sampled is homogeneous and the sampling technique reliably dispenses all particulates.

Data in Table 10-6 demonstrate that the robot system can conduct the digestion and that the results obtained agree well with the accepted analysis data.

Results from ICP analysis shown in Figures 10.6 and 10.7 also indicate that no apparent weight segregation occurs while operating the dispense-to-weight program. Another trial in which 12 samples of material with a known value of 3.28% copper and 4.80% zinc also confirmed that the samples weighed robotically are not segregated. Mean values for copper and zinc from robot-dispensed and digested samples were 3.28% and 4.77% respectively. Currently, each suite of samples processed by the robot system carries two quality control (QC) samples that are randomly placed within the suite of samples to be digested.

Discussion

Microwave dissolution can bring about enormous changes in the concept, construction, and utilization of the wet chemistry laboratory. The speed of microwave heating compared with electrical or gas-fired heating is remarkable. The usual acid-fume exhaust hood can be replaced by a less expensive fume extractor. The initial start-up costs or renovating expenses of a chem-

Table 10-6. Elemental Analysis of Tankhouse Anode Slimes After Robotic
Microwave Pressure Dissolution

Sample Number	Se(%)	Au(%)	As(%)	Sb(%)	Bi(%)	Te(%)	Pd(g/mg)
1	47.48	0.2504	0.55	0.34	0.70	0.05	36.84
2	47.96	0.2458	0.58	0.34	0.81	0.05	46.71
3	49.54	0.2590	0.64	0.37	0.83	0.06	39.95
4	47.32	0.2405	0.60	0.36	0.75	0.05	43.61
5	49.72	0.2447	0.66	0.38	0.78	0.06	38.39
6	50.14	0.2390	0.62	0.35	0.78	0.06	40.43
7	49.09	0.2549	0.61	0.37	0.86	0.06	42.79
8	51.76	0.2570	0.63	0.37	0.85	0.06	42.11
9	50.48	0.2461	0.62	0.36	0.78	0.06	42.96
10	50.93	0.2528	0.64	0.39	0.85	0.06	38.34
11	49.42	0.2437	0.62	0.36	0.86	0.06	41.49
12	50.24	0.2484	0.64	0.38	0.76	0.06	40.53
13	47.01	0.2489	0.59	0.32	0.78	0.05	39.10
14	48.53	0.2533	0.66	0.37	0.87	0.05	42.80
15	49.85	0.2657	0.62	0.36	0.80	0.05	39.75
16	49.58	0.2475	0.60	0.32	0.76	0.05	39.01
17	49.56	0.2522	0.57	0.37	0.82	0.05	41.64
18	47.01	0.2468	0.69	0.40	0.82	0.05	39.00
19	48.54	0.2536	0.65	0.39	0.76	0.06	40.04
20	49.91	0.2601	0.65	0.36	0.80	0.06	41.77
Mean	49.20	0.2510	0.62	0.36	0.80	0.055	40.86
V	1.753	0.000	0.001	0.000	0.002	0.000	5.195
SD	1.32	0.0070	0.034	0.021	0.045	0.005	2.279
CV	2.69	2.69	5.46	5.87	5.57	9.20	5.58
Accepted values	50.0	0.2517	0.64	0.37	0.80	0.06	41.1

NOTES: V is variance. SD is standard deviation. CV is coefficient of variation.

istry laboratory can be reduced. The use of manual microwave dissolution
can increase sample throughput and reduce analyst time in the laboratory.
It is no longer necessary to sweat over a hot plate to dissolve samples when
this clean and compact method of sample preparation is available. An added
benefit is the elimination of sodium peroxide fusion in nickel crucibles on
a semiroutine basis. Microwave dissolution using acids under pressure re-
places this tedious procedure.

The Robotically Controlled System

The most impressive achievement of this work is that, for the first time, an
ore sample can be routinely dissolved with mineral acids by a flexible robotic
system. The type of dissolution can be modified from the keyboard as required
by the user. Different programs can be written for varied applications using

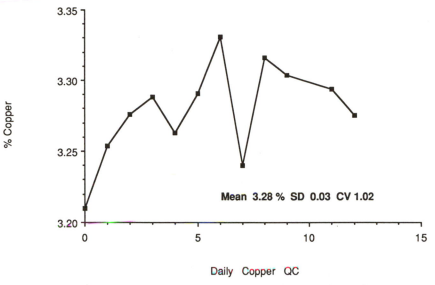

Figure 10.6. *Copper mill head quality control prepared by robot.*

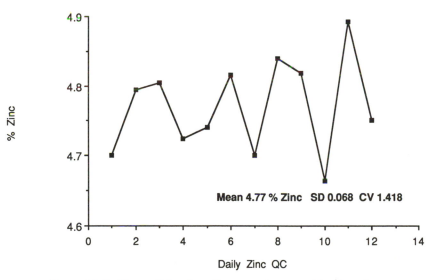

Figure 10.7. *Zinc mill head quality control prepared by the robot.*

the same system. The robot arm provides highly reproducible dissolution for six samples per hour. With an anticipated workday of 20 h, a total of 100 samples per day can be processed.

Another very positive aspect of the system is that it contributes to productivity without being a threat to job security. The system becomes an assistant to the team. Traditionally, chemical dissolutions were conducted in a laboratory that required a balance room, fume hoods, hot plates and adequate bench area to make up solutions to a final volume. With the automated system, the complete operation can be carried out on a 122- × 244-cm (4- × 8-ft) area equipped with three acid fume extractors. This concept in itself can significantly reduce the area required for future laboratories, as robots are already interfaced with AA and ICP units (8).

Mechanical Performance

The robot arm performed very well. Once the balance was stabilized and vibration-free, an excellent network of communication was established between the balance and the Zymate controller. The microwave heating unit and its door-opening operation have been trouble-free. Problems with the automatic vessel capper are being investigated. Modifications to correct mechanical and electrical failures due to corrosion caused by acid fumes are being implemented. The digestion vessel is now isolated within the capper, and all electronics have been made remote from the capper.

To remove acid fumes, an additional acid fume exhaust canopy has been installed. The master laboratory station is remote from the capper. The master laboratory station controlling valve for dispensing hydrochloric acid is also prone to failure and is undergoing evaluation. Preventative maintenance in this area is critical to successful implementation of an automated system.

System Support Requirements

Successful implementation of a robotics system requires a good support organization. The mere acquisition of a system does not guarantee that it will be successful. Staff for the project must be planned before acquiring the system, and should include the following:

1. Project manager, reporting to management and resources provider
2. System analyst, responsible for project concept
3. Computer programmer, to build the program step by step
4. Instrument technician, to deal with electronic failures and interfacing problems
5. Mechanical and electrical engineer, to design custom equipment
6. Analysts, willing to learn and create new technology

A good machine shop will also expedite the implementation of new ideas. This is a team project that requires the collaboration of several disciplines. The implementation of a robotically controlled pressure microwave dissolution system does not necessarily require a staff of six people; however, people with these qualifications must be available to assist the project manager when necessary. The project manager must keep an excellent communication network open. Robotic automation tends to have a high profile and is frequently misunderstood. The project manager's task is to inform the team and management of steps reached and problems encountered to achieve the overall objectives.

Because the system support will win or lose the game, this area is the most important. In its absence, the project should not even be considered. The idea of using a research chemist, a laboratory supervisor, or a laboratory analyst to "put the system together in their spare time" is a guarantee of failure. Analysts with a success-oriented attitude, creativity, and initiative are the secret to success. (9, 10). The analyst's attitude makes the robotic application successful.

Conclusion

The use of microwave technology for simple or complex sample dissolutions has proven itself to be a valuable technology available to meet increasing productivity demands when speed, precision, and accuracy are required. Robotically controlled microwave pressure dissolution is a natural evolution of manual microwave dissolution. However, we must live with the evolution from manual to robotic to fully appreciate the technical and social problems that arise. Our ultimate goal is to couple this robotic system to a flow injection unit that will feed an ICP. Chemistry is becoming a multifaceted career where computers and robots are performing the work conceived by the analyst. This chapter gives us a peek at the next century.

Acknowledgments

I thank Dennis Kemp for his approval to write this chapter and Len Green for his support. I also thank Richard Barrette for his dedicated efforts in serializing our in-lab program. The Zymark team of Mark Robinson and Warren Vollinger is acknowledged for its assistance in problem solving. Will Grooms and Dennis Manchester from CEM Corporation (Matthews, NC) and Douglas Biggs of Kidd Creek Mines Ltd. have contributed greatly to solving electronic problems. CEM's Steve Smith help with graphics is appreciated.

Appendix A: Dispense Sample-to-Weight Program

```
          VIB = 0.75
          WRIST=90
          TIMER  (1)=3
          WAIT FOR TIMER (1)
          WRIST=93
          DISPENSING.AT.SAVILLEX
   1 0    ? UNSTABLE. WT
          IF UNSTABLE .WT<0.150 THEN VIB=0.750
          IF UNSTABLE .WT<0.300 THEN VIB=0.550
          IF UNSTABLE.WT>0.300  THEN VIB=0.450
          IF UNSTABLE.WT <0.400 THEN VIB=0.250
          IF UNSTABLE.WT<0.400  THEN GOTO 10
          IF UNSTABLE.WT>=0.400 THEN VIB=0
     20   WRIST=86
          VIB=0
          OVER.SAVILLEX
          CLOSE.BALANCE.DOOR
     25   ? STABLE.WT
          IF STABLE.WT <0.400 THEN GOTO 30: ADD MORE SAMPLE
          IF STABLE .WT>0.400 THEN GOTO 40: VERIFY TOLERANCE.
```

Appendix B: Abort Program

```
       ? MAIN.POWER.ON
       IF MAIN.POWER.ON.=0 THEN 10
       IF MAIN.POWER.ON.=1 THEN 100
10     MICRO.POWER.OFF
       STOP.CAPPING
       STOP.UNCAPPING
       READY.TO.UNLOAD
       REMOVE.UNCAPPED.SAVILLEX
       S.DV.SENSING
       ? SENSOR
       IF SENSOR=0 THEN 20
       IF SENSOR=1.00 THEN 15
15     REMOVE.DV.FROM.ACID.DISPENSE
20      MOVE.TO.ABORT.STATION
       PLACE.ABORT.CAP.ON.SAVILLEX
       ABORT.A.AND.B.SYRINGES
       PARK.GP.HAND
       :TURN.ON.RED.ALARM
       STOP
100
```

3 5

Appendix C: Programs to Ensure Safe Operation

1. Safety micro switch to confirm that the balance door is open or closed.

```
          OPEN.BALANCE.DOOR  PROGRAM.
          BALANCE.DOOR.OPEN.FLAG=0
   10     BALANCE.DOOR.OPEN
          BALANCE.DOOR.OPEN.FLAG=BALANCE.DOOR.OPEN.FLAG+1
          IF BALANCE.DOOR.OPEN.FLAG=5 THEN STOP
          ? BALANCE.DOOR
           1.000000
          IF BALANCE.DOOR=0 THEN 10
```

2. Optical sensor to confirm the PFA vessel is present before acid addition and that the cap is present before capping and microwave heating.

```
          MOVE.SAVILLEX.TO.SENSOR PROGRAM.
          S.SAVILLEX FROM BALANCE
          S.SAVILLEX IN FRONT.OF.SENSOR
          S.SAVILLEX.SENSING
          SENSOR.ON.FLAG.=SENSOR.ON.FLAG+1
          ? SENSOR
           1.000000
          IF SENSOR.ON.FLAG=2 THEN STOP
          IF SENSOR=0 THEN PICK.UP.WEIGHED.SAVILLE
```

3. Flags to confirm that the microwave door is open.

```
      MICRO.DOOR.OPEN
      MICRO.DOOR.FLAG=0
      MICRO.DOOR.CLOSE.FLAG=0
      MICRO.POWER.OFF
10    OPEN.MICRO.DOOR
      TIMER (1)=5
      WAIT FOR TIMER (1)
      MICRO.DOOR.FLAG=MICRO.DOOR.FLAG+1
      ? MICROWAVE.DOOR
      1.000000
      IF MICRO.DOOR.FLAG=5 THEN STOP
```

4. Tactiles sensing to confirm Savillex is free and not jammed into the capper

```
      REMOVE. CAPPED. SAVILLEX
      OVER.CEM.CAPPER
20    GRIP=200
      DOWN.TO.CEM.CAPPER WITH.SAVILLEX
      GRIP=10
      ? GRIPFORCE
      0.750
      IF GRIPFORCE<0.5 THEN GOTO 20: TRY AGAIN
      IF GRIPFORCE>0.5 THEN GOTO 30: CONTINUE
30    OVER.CEM.CAPPER
```

Literature Cited

1. Matthes, S. A.; Farrell, F. R.; Mackie, A. J. *Bureau of Mines, Technical Progress Report 120,* **1983,** 9.
2. Merks, J. W.; *Sampling and Weighing of Bulk Solids 1985;* Trans Tech, Karl Distributors, Rockport, MA 01966; Vol. 4; p 242.
3. Labrecque, J. M. *Proceedings of the 17th Annual Canadian Mineral Analysts Conference,* September 1985, p 17-38.
4. Sutarno, R.; Steger, H. F. *Talanta,* **1985,** *32*(6), 439-445.
5. Sample, Robert M. *Proceedings of the Analytical Division of the Royal Society of Chemistry* July 1986, Vol. 23, No.7.
6. *Zymate System Instruction Manual;* Zymark: Hopkinton, MA, 1983.
7. Pribil, R. *Analytical Applications of EDTA and Related Compounds 1972;* Belcher, R.; Freiser, H. Eds.; Pergamon: New York Vol. 52; p 83.
8. Lester, L.; Lincoln, T.; Donoian, H. *Advances in Laboratory Automation Robotics 1985;* Strimaitis, J. R., Hawk G.L., Eds.; Zymark: Hopkinton, MA, 1985; Vol. 2; p 509.
9. Miller, Robert G. *Advances in Laboratory Automation Robotics 1985;* Strimaitis, J. R.; Hawk, G. L., Eds.; Zymark: Hopkinton, MA, 1985; Vol. 2; p 31.
10. Zenie, Francis H. *Advances in Laboratory Automation Robotics 1985;* Strimaitis, J. R.; Hawk, G. L., Eds.; Zymark: Hopkinton, MA 1985; Vol. 2; p 43.

RECEIVED for review September 30, 1987. ACCEPTED revised manuscript March 11, 1988.

CHAPTER 11

Safety Guidelines for Microwave Systems in the Analytical Laboratory

H. M. Kingston and L. B. Jassie

"It is the characteristic of wisdom not to do desperate things".

Henry David Thoreau

Safety considerations for working with microwave systems in the laboratory are discussed. The regulatory implications of modifying microwave equipment are presented. Guidelines for the selection and use of common laboratory equipment and materials are examined.

SAFETY, ALWAYS AN IMPORTANT CONSIDERATION in analytical chemistry, is especially important during the development of new procedures. The relatively new technique of microwave acid decomposition or dissolution in open or closed vessels introduces unique factors that many analysts have not encountered. Acid combinations traditionally used to digest a particular matrix may be inappropriate when microwave energy is used as the heat source. Differences between traditional acid dissolution and microwave acid dissolution should be examined before using microwave energy to heat acids or samples directly.

The use of strong mineral acids to decompose samples before elemental determination is a well-established practice in analytical chemistry (1). This method has been used for all types of samples, from geological, (2, 3) biological, and botanical samples to alloys (3). The combinations of acids, reagents, samples, and methods that are safe and reliable are not well compiled. Accepted procedures and good laboratory practices are scattered throughout the literature, as are the recommended safety suggestions for using strong acids to decompose samples. Complete coverage of the literature and of the range of possible sample types is beyond the scope of this chapter. Rather, the new subject of safety for microwave acid decomposition will be considered here. Information on safe combinations of acids, reagents, and

1450–6/88/0231$6.00/0

samples can be found in chemical safety literature such as *Dangerous Properties of Industrial Materials* by Sax (4); works devoted to the decomposition of samples, such as *A Handbook of Decomposition Methods in Analytical Chemistry* by Bock (3); and extensive studies of individual acids, such as *Perchloric Acid and Perchlorates* by Schilt (2).

Microwave Energy

Microwave energy is not only absorbed by the mineral acids to produce heat in the dissolution of samples, but is also absorbed by some sample molecules, and this absorption increases the kinetic energy of the matrix. The interactions of radiation with molecules can be inconsequential or extremely significant, depending on the composition and amount of material present. In general, polar and ionic molecules interact with microwave energy through the mechanisms of dipole rotation and ionic conductance. These mechanisms are described in detail in Chapter 2. These interactions may aid the decomposition of some samples and partially explain the rapid decompositions that often appear to be faster than can be explained solely by the increased heating rate of the acid. Microwave energy can be projected over distances and propagated through materials using different mechanisms than more conventional heating methods. Hence, the use of microwave energy in sample decomposition should not be regarded as simply a way of heating acid.

Microwave Energy Leakage Standards and Regulations

Currently, microwave leakage from ovens intended for heating or cooking food is limited to 5 mW/cm^2 at a distance of 5 cm from any surface of the oven or from an insulated wire inserted through any hole into an energy-containing space. These performance standards are incorporated in the Radiation Control for Health and Safety Act, a Federal Law enacted in 1968. The regulations were promulgated in 1970 and the standard became effective in 1971 and is administered by the Center for Devices and Radiological Health (CDRH) of the U.S. Food and Drug Administration (FDA). The regulations are contained in Title 21 of the Code of Federal Regulations (CFR), Part 1030.10. Although there is no microwave leakage limit for laboratory and scientific units, manufacturers of these products are subject to other FDA regulations (21 CFR 1002–1004), such as: reports[*] of design and quality control, including radiation safety measures; reports[*] of acci-

---

[*] Such reports should be addressed to FDA, Center for Devices and Radiological Health, 5600 Fishers Lane, HFZ-312, Rockville, MD 20857.

dental radiation exposures to users or service or production personnel; and recall* of any product that is found defective (i.e., presents a risk of radiation injury to any person). Manufacturers subject to the FDA regulations include original manufacturers, importers, and persons who remanufacture products for distribution to others. Remanufacturing includes adapting a product for a new intended use (such as converting household cooking ovens for laboratory use) and reselling them, but it does not include user modification of his or her own equipment.

For laboratory, industrial, and medical products, the Occupational Safety and Health Administration (OSHA) requires that the maximum exposure to radio frequency (rf) energy for an operator in a safe work place is <10 mW/cm^2 averaged over 6 min (established in 1970 in 29 CFR 1910.97). OSHA regulations also require that if microwave equipment is modified or the integrity of a safety device is violated, the product must be demonstrated to be safe by measuring the microwave radiation exposure potential. Twenty-nine states have their own regulations, and some have adopted the latest American National Standards Institute (ANSI) guideline of 5-mW/cm^2 maximum exposure at a frequency of 2450 MHz.

Safety Devices in Microwave Units

In addition to energy leakage standards, safety interlocking devices are required on all commercial and consumer microwave equipment to prevent accidental exposure (5). These devices should not be removed or defeated on equipment used in the laboratory. They protect against initiation or continuation of the production of microwave energy if the cavity is not sealed. Other safety devices, such as the closed waveguide and door seals, should be inspected if corrosion becomes apparent. *Placing a microwave system in a hood, where it is surrounded by acid fumes, causes accelerated corrosion of the equipment and should be avoided.* Fumes should be transported away from the unit or the cavity air swept away to a hood. Deterioration of the waveguide, door seals, or cavity walls could allow the escape of microwave energy. These events cannot be prevented by interlocking devices, but can only be controlled if the analyst frequently checks the system to ensure that the equipment is maintained in safe working order. Also, if any portion of the microwave unit or door seal becomes damaged by a catastrophic event such as an acid spill or impact, the safety of the equipment should be reevaluated before it is put back in use. Moderately priced microwave survey instruments are currently available to detect microwave leakage (*see* Appendix A) (6). Spacer cones should always be used to protect the sensitive detecting elements and should be replaced when worn or dirty to avoid additional measurement errors. Survey instrument calibration should be checked and recalibration performed by qualified personnel.

Modifying Microwave Equipment

Microwave equipment sometimes must be modified to produce new configurations that permit the direct addition of reagents or the measurement of temperature and pressure in the microwave cavity. These modifications should be undertaken only by those trained in high-voltage and microwave equipment. After any modification of the equipment, it is important to ensure that no path has been created for radiation to leave the internal cavity. Measurements to detect such microwave radiation should be performed at and around the modified portion of the equipment. A microwave survey meter with good sensitivity is the recommended for testing alterations for leakage. The unit should be run at full power with no load, or if an appliance-grade microwave oven is being evaluated, a small load (50–100 mL of water) should be in the unit at the time of testing to prevent damage to the magnetron.

Microwave equipment can be modified by attachment of a wavelength attenuator cutoff to conduct tubing, thermocouple wires, and fiber optics into the microwave cavity. These devices have been made in many configurations. Two types of attenuators that we have tested are rigid stainless steel tubing (7) and flexible, tin-plated copper braid (8). The common construction features of both attenuators are metallic tubing of a conducting metal with the smallest diameter possible that allows the tube, wire, or device to pass through. This cutoff device is effective only if the diameter is small and the length is long compared to the wavelength of the 2450-MHz radiation (which is approximately 12.25 cm long). Dimensions of the devices that we used were approximately 0.7 cm i.d. and between 35 and 70 cm long. The attenuator must be grounded by making contact with the microwave cavity wall around the hole. Metallic devices passing through the attenuator must be grounded at the microwave cavity wall to prevent the wire or metallic tube from acting as an antenna to transport the electromagnetic energy out of the cavity and totally defeating the effectiveness of the wavelength attenuator cutoff. Experiments in which thermocouple wires were passed through the attenuator without shield and grounding were carried out, and energy fields >10 mW/cm^2 were measured at the end of the attenuator. Six separate attenuators were tested and found to prevent microwave radiation from escaping (<0.01 mW/cm^2 was the limit of our detection equipment). The inside hole in the cavity wall should be as small as possible. A small hole is also an effective barrier to microwave energy, as can be seen by observing any microwave cavity at the air intakes or through the grid on the inside of a transparent microwave door.

Problem Areas and Potential Difficulties

Problems have been encountered by researchers working on microwave acid decompositions. In many cases, these experiences were not reported in the literature but were shared with other researchers at scientific meetings or in personal conversations.

Vessel Failures

Overpressurization of vessels was common in the early work with closed containers. The failure is related to the material and design of the vessel, and although almost any vessel can be overpressurized if not fitted with a pressure-relief device, different types of vessels have very different modes of failure.

Many catastrophic failures of closed fluorocarbon vessels have been in those made from heavy-walled poly(tetrafluoroethylene) (Teflon PTFE). Pictures of vessels sheared in half and with threads stripped away from the vessel body have been circulated among researchers. Different fluorocarbon polymers have very different physical characteristics. In tests in which the Teflon PFA vessels (Savillex, 120-mL) were stressed beyond their ability to contain the pressure, the failure mode was not catastrophic. These vessels relieved pressure either by cracking and venting the internal gases or by losing the seal between the cap and container body.

Polycarbonate bottles, which were among the first closed containers used, had a 5–10% failure rate and were replaced by Teflon PFA vessels. Containers that have not been thoroughly evaluated for high-pressure applications should not be used unless the manufacturer has designed and tested them for this environment. Unless pressure and temperature measurements are made in real time, a pressure-relief device should be used to protect against overstressing the container, even if it has been tested for elevated temperature and pressure applications.

Equipment Failure

Commercial microwave appliances that are intended for cooking food are not designed to withstand chemical attack from corrosive substances. One accident has already occurred (case 18-138 reported to Microwave/Acoustic Products Section, FDA, 1986) in which all the safety-interlocking devices were rendered inoperative as the result of chemical interaction with the metal switches. An operator was exposed to the full power of the microwave unit for approximately 1 min. This is the first reported case of human injury from the use of a home microwave unit in the laboratory. These units were not designed for use in a chemical environment, and interactions between the instrumentation and chemicals should be considered before equipment is purchased. Corrosive environments, especially acid vapors, should be avoided and precautions taken to keep the appliance-grade equipment safe from chemical attack.

Sparking and Metallic Samples

Many alloys interact with microwave energy and become hot enough to melt the plastic vessels; other alloys accumulate large electrical potentials. Alloy samples can melt through the container, and several types of ferrous alloys have been tested in which spark discharges were quite spectacular. Large pieces of metallic samples should be avoided, because electrical arcs may be formed between individual sample pieces or between alloys and the microwave cavity walls or floor. The formation and intensity of electrical sparks depend on the composition of the alloy and other conditions such as field strength. These sparks may be energetic enough to puncture the sample vessel. Electrical arcs from inadequately grounded thermocouples constructed of no. 316 stainless steel were sufficiently energetic to puncture 1/16-in. Teflon PFA tubing wall (9).

The mineral acid decomposition of metal and alloy samples has a unique hazard that must be considered before traditional methods of hot mineral acid decomposition are applied to the microwave environment. Metals below hydrogen in the electromotive series readily liberate hydrogen when dissolved in acid. A potentially flammable or explosive mixture with oxygen may result if hydrogen is formed, and if the sample is in an open beaker or its closed container is sealed in air. In traditional digestion procedures an ignition source would not be present, but in microwave digestion metallic particles can interact with the strong electromagnetic field, a spark may be generated from the sample, and ignition of the hydrogen may cause a fire or explosion.

To prevent this potentially hazardous situation, closed digestion systems should be sealed in inert gas atmospheres. In open-beaker applications, purging the compartment with inert gases will keep air from the sample. The limits of flammability of hydrogen in air diluted with an inert gas were presented by Lewis et al. (10). In open-vessel digestions, rapid removal of hydrogen may also prevent the formation of the hydrogen–oxygen mixture necessary for ignition. Unfortunately, the amount of hydrogen in air necessary for burning is from 4 to 75% (11) and limits of flammability of hydrogen in oxygen are from 4 to 94% (10). The energy of activation necessary to ignite this mixture can be achieved by the weakest spark, or the mixture may be catalytically ignited. Therefore, the production of hydrogen and its mixtures with oxygen anywhere in this range are extremely hazardous.

Flammable Solvents

Another hazard that may become more prevalent as chemical applications for microwave equipment diversify is the use of flammable solvents in the electromagnetic environment. Such a problem could have occurred when students heated diethyl ether in an open container (12). That occasion was without mishap. However, because sparks are common in microwave systems, flammable or explosive substances mixed with oxygen from the air pose especially hazardous situations.

Predicting Digestion Conditions

By using the fundamental equation relating power consumption of microwaves by acids [described in Chapter 6 and in the literature (7)], the analyst can predict the actual conditions in some digestion media and vessels and estimate the conditions in others accurately enough to prevent dangerous situations from occurring. It is imperative that the analyst knows the temperature limits of the vessel material to prevent thermal degradation by overheating and uses established pressure and temperature limits for commonly used microwave digestion vessels as a guideline to provide the upper-limit targets for these calculations. New vessel designs and new materials will continue to be introduced, and common sense must prevail in evaluating the pressure and temperature stress that may be applied to untested containers.

Exothermic reactions have been documented (6) when extreme power settings were used with oxidizing acids and organic materials. When planning decomposition conditions, the analyst must consider that the oxidizing power of an acid may increase with temperature and that exothermic reactions may occur.

Increased Reaction Rates

Acid temperatures reached in 5–30 min by traditional methods are achieved in seconds in the microwave cavity, and reaction times are greatly reduced. The coupling of electromagnetic energy with certain molecular dipoles in the sample may aid the digestion and further accelerate the reaction. With rapid heating, reaction products may be evolved too quickly and in quantities too large to be vented or contained. Thus, the reaction should not be accelerated to such an extent that the safety of the individual or structural limits of the equipment will be compromised.

Microwave Field Variability

The microwave energy field is not homogeneous enough to allow the placement of samples at fixed positions in the cavity and achieve reproducible, uniform energy exposure. Even when mode stirrers are used to homogenize the field, temperature differences (°C) of as much as 50% can be observed in identical vessels placed at different locations within the microwave cavity. To provide the same digestion conditions for all samples, and to ensure that the vessels are kept within temperature and pressure limits, it is necessary to maintain a uniform energy exposure of all sample containers. Rotation of the containers through the energy field minimizes exposure differences between samples. As seen in Table 11-1, reproducible temperatures (within 1 to 3%) have been achieved over the range 23–180 °C with a rotation of 360° every 20 s about a 12-cm radius in a 600-W field.

Laboratory Vessels in the Microwave Environment

Sample preparations, organic synthesis, and other laboratory operations using microwave energy require many different types of sample vessels. The suitability of the vessel material depends on the specific conditions, use, and solvent for which the vessel is intended. When choosing a container material for open-vessel work, the boiling point of the acid or solvent being used must be known because it is the maximal temperature to which the material will be exposed. When closed systems are used, temperature measurement is necessary to prevent the liquid contents of the vessel from exceeding the upper temperature limit of the vessel material, which is normally the maximal continuous service temperature. Below this temperature, a polymeric device will continue to function as intended, without change, for an extended period of time. Above this temperature, polymeric materials may deform, deteriorate, or lose some essential property so that they no longer function optimally. For example, a high-boiling solvent like sulfuric acid will melt Teflon but not borosilicate glass or quartz.

Closed vessels are subject to stress as the result of the partial pressures of the heated solvents and decomposition products. Excess pressure can be controlled by appropriate venting. Pressure can be monitored and regulated

Table 11-1. Temperature Reproducibility with Microwave Heating

Time (min)	Group 1 Temperature (°C)	Pressure (atm)	Group 2 Temperature (°C)	Pressure (atm)
0	23	0.99	25	1.00
1	84	1.10	78	1.08
2	110	1.31	114	1.32
3	126	1.57	125	1.53
4	130	1.74	133	1.88
5	138	2.18	136	2.18
6	151	3.48	152	3.24
7	164	4.69	162	4.49
8	170	6.20	172	5.63
9	172	7.10	178	6.96
10	178	8.40	181	7.65

NOTE: Measurement of one sample in two separate batches of six 1-g rice flour samples each in 10 mL of nitric acid. Two-stage heating program: 402 W for 5 min then 574 W for 5 min.

by controlling the temperature of the solvent or limiting the sample size to restrict the quantity of gaseous decomposition products. Pressure that can be contained at room temperature may be too high when the vessel is at an elevated temperature. This condition is especially true for polymeric materials. Normal laboratory glassware should not be used for closed-vessel digestions and pressurized-glass containers have not yet been demonstrated safe for microwave digestions.

After chemical and mechanical considerations, the dielectric constant and microwave energy absorption characteristics of the material are important physical factors to consider. Many materials can be used in a microwave field because they do not absorb the energy. Among the many polymers that are transparent to microwave radiation are all types of fluorocarbons, especially Teflon PFA, which is ideal for use as a sample container, because of its exceptional chemical and thermal durability. Other common laboratory polymers, such as polyethylene and polypropylene, are also suitable for use in the microwave environment.

When choosing containers for use in a microwave field, the composition of all parts needs to be considered (e.g., handles or screw-on caps, which tend to be made from different materials than the vessel itself). Problems can be avoided by choosing materials that do not significantly absorb microwave radiation or absorb very little. Listed in Table 11-2 are the dielectric constants, melting points, and (maximum) continuous use temperatures of some common laboratory container materials. Those with small dielectric constants at 25 °C tend to be good electric insulators and exhibit little or no heating in the microwave environment. Data on dielectric constants at microwave frequencies (1×10^9 Hz) are sparse, but when known (13), are not very different from values at the more common frequency of 10^6 Hz used in Table 11-2. Thus, fluorinated polymers are excellent materials for microwave containers because of their low dielectric constant, and are essentially transparent at 2450 MHz. Laboratory vessels fabricated from or incorporating Bakelite, Lucite, or other thermoplastic resins may be acceptable, even though they absorb a small amount of energy. A material that does not appear on the list and whose electromagnetic properties are unknown can be tested by placing it in a microwave unit for 30–90 s at full power to determine whether it absorbs energy, as evidenced by a measurable increase in temperature.

Other materials that can be used in a microwave environment include fiberglass and polyurethane foams found frequently as vessel insulation or lining. If homemade racks, carousels, or sample holders will be used, some thought must be given to construction materials to ensure compatibility. Ordinary polyethylene or Teflon laboratory tubing can be used safely in the cavity for handling gas.

Metals and metallic products are generally unsuitable for use in the microwave environment because they reflect electromagnetic radiation, form

Table 11-2. Thermal and Microwave Characteristics of Laboratory Container Materials

Material	Melting Point (°C)	Maximum Service Temperature (°C)	Tangent $\partial \times 10^4$
Teflon, PFA	302	260	2.1
Teflon, FEP	252–262	204	2.1
Halon, PTFE	>320	260	2.1
Poly(methylpentene)	240[a]	175[b]	2.1
Kel-F, CTFE	198–211	199	2.3–2.4
Tefzel, TFE+CE	271	200	2.6
Polyethylene	120–135	71–93	2.2–2.3
Polyethylene(HD)	146[b]	121[c]	2.25
Polypropylene	168–171	100–105	2.24–2.4
Poly(methyl methacrylate)	115[b]	76–88[b]	2.6
Polystyrene	242[b]	82–91	2.7–3.1
Polycarbonate	241	121	2.9
Poly(vinyl chloride)	>199	60	3.2–3.3
Polysulfone	<190[d]	160[d]	3.0
Nylon 6	216	102	3.4
Nylon 6/10	>215	80–102	3.1–3.5
Nylon 6/12	212	102	3.5
Nylon 6/6	253	102	3.6–3.7
Polyacetal, copolymer	165–175	121	3.7
Glass, quartz	>1665		3.8–4.1
Bakelite, (asb. fill)	decomposes	200–218	3.7–4.8
Phenol/Formaldehyde	decomposes	120–190	4.1–5.0
Glass, borosilicate	>1080		6.3–6.8
Ceramic (depends on type)			6.5
Porcelain (depends on type)			6.0–8.0

Abbreviations: PFA, fully fluorinated, long-chain carbon polymer with pendent perfluoroalkoxy side chains; FEP, fluorinated ethylene polymer; PTFE, (poly)tetrafluoroethylene; and CTFE, chlorinated tetrafluoroethylene.
NOTE: Microwave absorption increased from Teflon PFA to porcelain. Materials with larger dielectric constants will heat more in the field.
SOURCE: Adapted from Ref. 14.
[a]Ref. 15.
[b]Ref. 16.
[c]Ref. 17.
[d]Ref. 18.

intense hot spots in the cavity, and can accumulate an electric charge powerful enough to arc weld metal. If they are used, they must, like thermistors and thermocouple probes, be shielded and grounded for use in the cavity.

Conclusion

The safety of the analyst is the most important consideration in the analytical process and in sample preparation. As the result of direct coupling, a great deal of microwave power is available. The advantages of high temperature are quickly realized, but control of sample heating can be lost just as quickly. As in all good laboratory practices, there is no substitute for common sense. Common sense dictates careful planning and cautious experimentation when the results are uncertain and the equipment is unfamiliar. Most of the mishaps reported have shown lapses in common sense.

Occasionally, an unpredictable event will result in equipment damage and loss of the sample. These incidents should be reported as soon as they are discovered to help others avoid these same difficulties. Because science progresses by virtue of the observations and discoveries of our predecessors and peers, we can further our art by making these careful observations known. That is the purpose of the scientific literature. As this collection of observations becomes obsolete, newer observations will continue to advance this field as a science. We, the authors of this volume, hope we have provided some insight into what has been studied, and anxiously await the new contributions to this subject that will be made.

Appendix A: Hazard Monitors for the Detection of Microwave Leakage

The following companies sell moderately priced microwave survey instruments.

Anchor Chemical Australia Pty., Ltd.
Box 474
P. O. Crow's Nest
N.S.W. 2065
Australia
(02) 439-2144

Applied Microwave Energy, Inc.
31127 Via Colinas
Westlake Village, CA 91362
(213) 991-4624

Bach–Simpson, Ltd.
1255 Brydges Street London,
Ontario N5W 2C2
Canada
(519) 452-3200

General Microwave Corporation
155 Marine Street
Farmingdale, NY 11735
(516) 694-3600

Gerling Laboratories
1628 Kansas Ave.
Modesto, CA 95351
(209) 521-6549

Holaday Industries, Inc.
14825 Martin Drive
Eden Prarie, MN 55344
(612) 934-4920

Micor, Inc.
3901 Westerly Place, Suite 102
Newport Beach, CA 92660
(714) 476-0616

Microwave Heating, Ltd.
1A Heron Trading Estate
Luton Beds, England LU3 3BB
United Kingdom
(0582) 58474

Narda Microwave Corporation
435 Moreland Road
Happauge, NY 11788
(516) 231-1700

Literature Cited

1. Hillebrand, W. F.; Lundell, G. E. F.; Bright, H. A.; *Applied Inorganic Analysis*; Wiley: New York, 1953; 2nd ed.
2. Schilt, A. A. *Perchloric Acid and Perchlorates*; G. Frederic Smith Chemical: Columbus, OH, 1979.
3. Bock, R. *A Handbook of Decomposition Methods in Analytical Chemistry*; translated and revised by Marr, I. L.; Wiley: New York, 1979; Chapter 4.
4. Sax, N. I. *Dangerous Properties of Industrial Materials*; Van Nostrand Reinhold: New York, 1979; 5th ed.
5. Copson, D. A. *Microwave Heating*; Avi: Westport, CT, 1975; p 443.
6. *International Directory of Electromagnetic Heating and Instrumentation 1986–1987*; Wyslouzil, W.; Ed., International Microwave Power Institute: Vienna, VA, 1986.
7. Kingston, H. M.; Jassie, L. B. *Anal. Chem.* **1986,** *58*, 2534–2541.
8. Kingston, H. M.; Jassie, L. B. Presented at the 25th Eastern Analytical Symposium, New York, October, 1986, paper 76.
9. Kingston, H. M.; Jassie, L. B.; Fassett, J. D., Presented at the 190th National Meeting of American Chemical Society, Chicago, IL, September, 1985, paper ANYL 10.
10. Lewis, B.; Guenther, E. *Combustion, Flames and Explosions of Gases*; Academic: New York, 1961; 2nd ed.; p 695.
11. *Hazardous Chemicals Data Book*; Wise, G., Ed.; Noyes Data: Park Ridge, NJ, 1980.
12. Bedson, A. *Chemistry in Britain* **1986,** *22*(10), 894.
13. *Dielectric Materials and Applications*; Von Hippel, A. R., Ed.; Technology Press of M.I.T. and Wiley: New York, 1954; p 301.
14. *Plastics 1980*; Gosnell, R.; Kusy, P. F.; Keimel, F. A.; Miller, H.L.; Pebly, H. E., Eds.; International Plastics Selector: San Diego, CA, 1979.

15. *Handbook of Plastics and Elastomers*; Harper, C. A., Ed.; McGraw–Hill: New York, 1975.
16. *User's Practical Selection Handbook for Optimum Plastics, Rubbers and Adhesives*; ITI: Tokyo, Japan, 1976.
17. *Polymethylpentene "TPX"*, Mitsui Petrochemical Industries, Ltd., Kasumig Aseki 3-Chome; Chiyoda-ku: Tokyo 100, Japan.
18. *Modern Plastics Encylcopedia 1988*; Juran, R., Ed.; McGraw–Hill: New York, 1987.

RECEIVED for review June 30, 1987. ACCEPTED revised manuscript February 11, 1988.

Index

Index

A

Copyediting and indexing by Linda Romaine Ross
Production by Barbara J. Libengood
Managing Editor Janet S. Dodd

Elements typeset by Techna Type, Inc., York, PA
Printed and bound by Maple Press, York, PA

Recent ACS Books